Garden Sass

Garden Sass

The Story of Vegetables

by

Lucile McDonald

THOMAS NELSON INC.

New York / Camden

Library of Congress Catalog Card Number: 77-145921
International Standard Book Number: 0-8407-6124-4
Manufactured in the United States of America

Contents

1.

Man and Hunger

From the time people were first created they were hungry, and long before written history began man became an agriculturist. Every principal vegetable food used today was already domesticated by the time the first picture writing was chiseled on stones. Therefore we do not know exactly by what means early man tamed and improved the wild onions, beans, cucumbers, and cabbage he found growing around him. We can only guess that ancient farmers saved the best of their seeds each season to plant the following spring, and that in this way they gradually gave to the world families of superior food plants, quite unlike the weeds to which they were closely related.

Earliest man possessed no weapons except the rough stones he found on the ground or the branches he broke from trees. He lived poorly, never knowing for certain where his next meal would come from. He wandered on the prairie and in the forest, seeking whatever was ripe —sometimes seeds, fruits, and nuts, sometimes tender young shoots, sometimes mushrooms or succulent leaves. When these were scarce, he contented himself with edible roots and with grubs and other insects.

One day as he reached home with his gleanings, one of these ancient inhabitants of the earth either accidentally dropped or purposely scattered a few seeds in the clearing around his dwelling. The next spring he found tender shoots pushing out of the soil where the seeds had fallen. As they thrived and bore fruit, the man must have realized that he could grow his own favorite foods. And in this simple act of an individual, agriculture probably had its beginning.

This discovery altered the habits of every race. From the time men knew how plants could be grown, they sowed seeds in the rough clearings around their homes. By the New Stone Age, man had created tools suited to the cultivation of plants. Gradually he felt the desire to settle in one place and adapt himself to a new way of life. Before long he noticed that some of the food plants in his crude garden were much better flavored than their wild cousins.

Before the beginning of agriculture there was constant warfare among primitive people to decide who should possess good hunting grounds and areas where fruits, seeds, roots, and herbs grew in greatest abundance. Groups organized to protect the land on which they gathered their food, and in this way tribes were formed. As soon as the increasing interest in agriculture caused men to settle down and grow what they ate, there was less fighting and some of the warriors turned their attention to other crafts.

When man became a food producer instead of a food gatherer, groups of wanderers consolidated into villages and stayed at home to cultivate crops and take care of their domestic animals. When they at last had plenty to eat and more time to spare, they made larger and more permanent settlements, and so cities were born. Industries began and the whole pattern of life was changed.

But agriculture did not spring up just anywhere on the globe. Intelligent tribes settled where wild plants were most plentiful. In nature's primeval garden lands, well-fed people thought more clearly, learned more quickly, and developed arts and crafts more freely than underfed persons living in barren places. Because intelligent tribes settled in spots where plants were plentiful, almost every vegetable we eat today came from the birthplace of some great civilization. In the Old World these regions were the Middle East, the lands surrounding the Mediterranean, and the mountainous parts of China; in the New World they were Mexico and Peru.

After a time, large stretches of land were covered with gardens and fields. When tribes needed more room they moved to new locations. This went on for thousands of years and still is going on. It meant more than a relocation of people; it was also a migration of plants. Many seeds that man did not carry intentionally clung to his clothing or to the hairy hides of domestic animals and dropped off along the way. Seeds sometimes were carried by marine currents, wafted by the wind, or transported in the undigested food of migrant birds. Gradually, by one means or another, every country has come to have among its fruits, grains, and vegetables many that were once strange to it.

Had he been better armed, the caveman might have brought home more meat. However, he could not spend precious daylight hours fashioning a stone spear tip if his stomach was empty. Each morning he was forced to halt whatever clumsy work he was doing and gather the seeds and berries that kept him alive. As his intelligence and skill increased, he invented snares, stone axes, fishhooks, and arrows. These enabled him to obtain different kinds of food with less labor, but everything he could hunt or fish or gather he ate raw.

Finally, when prehistoric man discovered the use of fire, his method of eating changed. He noticed that meat that had been near the coals tasted more savory than the raw flesh to which he was accustomed. Very likely he owed this discovery to a forest fire. Hunger may have driven him to sample the cooked flesh of an animal trapped by the blaze. Finding the meat more tender and better flavored, he probably decided to try cooking some himself. He may also have held bundles of grain over the fire, burned off the chaff, and munched the parched kernels. Next, he doubtless placed vegetables near the embers or wrapped them in leaves and laid them on the coals. Observing how cooking improved their taste, he devised underground ovens. In time he learned to boil water in a gourd or shell and cooked his food in such vessels or in clay pots that he fashioned.

When man had reached this stage in his development, he was able to consume coarse vegetables of the leafy type in greater variety and larger amounts than when he had to eat them raw. He no longer had to spend his time seeking only those he could eat uncooked. He now had added cooked game and fish to his diet. Sometimes, when partridges were scarce or wild boar hid deep in the forest or trout were not tempted by his line, he was forced to return home empty-handed. But after he had found that cooking could make edible many things he had never before been able to eat, he had a more certain way to keep his clay pots filled with food for his hungry children. The surplus, which he spread on the ground to dry in the sun, could be used in winter or in times of scarcity.

Since that early period, man has improved food plants and increased the varieties; today there are many thousands. He now habitually eats those produced by cultivation. Even though roots and fern shoots which Indians once sought still abound in rural areas, modern man would not think of gathering them except in extreme necessity. Today if we add a new dish to our diet, we borrow from the cultivated plants of other nations.

Although scientists speak of a time when, because of the population explosion, we may have to seek sources of nutriments from tree leaves and grasses and from vegetation in the sea, that day is in the far future. Production of foods with which we are already familiar can be expanded manyfold, using our present equipment and know-how. Much arable land still lies idle in the world. Giant machines have been developed that can do the work of scores of men who once were required to raise and harvest the variety and abundance of farm foods that now reach our dinner tables the year around.

In this book the items referred to as vegetables are not confined strictly to the botanical definition of that word, which applies only to roots, bulbs, tubers, shoots, and the stalks or leaves of plants used for food. Scientifically speaking, corn is a cereal, and tomatoes, eggplant, peppers, peas, and squashes are all fruits, for they are the seed containers of plants. Artichokes and cauliflower are portions of flowers. However, we shall refer to as "vegetables" all the things so called in grocery stores and at family tables.

These foods are so familiar that we cannot think of them as different from their present state. We cannot regard tomatoes and eggplant as poison or remember when there was no broccoli. It does not occur to us that peas, celery, potatoes, and cabbage did not always taste so good. Man has made many changes in the plants which, in early days, he found growing wild around him, and these changes are still going on.

2.

Peas and Lentils

Seeds must have been among the first things man nibbled in his early quests for food. He chose those that were most plentiful and sustaining, among them peas, beans, and lentils. Because these three vines of the Leguminosae family almost always bore a plentiful yield, primitive farmers included them in their earliest plantings. They knew that the pods ripened in a short season and did not require much care.

During long periods when game was scarce and meat and fats were lacking, whole races might have perished if man had not been wise enough to eat these nourishing seeds. He did not know that their richness in proteins made them valuable body-building substances; he thought of them only as food for man and beast, which could be consumed at once or stored for several years.

Ancient historians do not say which were first eaten, for those early writers hardly knew the difference between peas, beans, and lentils. All could be used in pottage, the principal cooked food of antiquity. This was a cross between thick soup and porridge, simmered many hours in a clay jar set on the coals, sometimes with meat or herbs added for flavoring.

Until several hundred years ago, only common field peas, which are fed to cattle, were known to man. These may have been eaten first in Ethiopia or Asia Minor, Turkestan, or India. We can be sure from seeds found in their ruins that the lake dwellers of Switzerland had peas in the Old Stone Age. They were also common in Egypt and Israel. The Bible relates that peas, lentils, and beans were among the foods brought to David in the desert. "[They] brought wheat, barley, meal, parched grain, beans and lentils . . ." (II Sam. 17:28.)

Lentils come from a leguminous plant with pale-purple flowers and short, flat pods. The Israelites and Egyptians ground the seeds between stones and made a flour for bread. Such loaves were offered to the dead at Egyptian funerals.

Peas or lentils used in soup are cooked a long time to make their coarse cell walls digestible. Thick soup, somewhat like the pottage of old, is still the most common dish made from lentils.

A famous story in the Bible tells about the twin brothers Esau and Jacob. Esau was a skilled hunter; his brother was a hard-working plodder who remained at home and farmed. One day Esau went to Jacob's tent, faint from hunger, and pleaded for some lentil pottage.

Jacob disapproved of the way in which Esau spent his time and was unwilling to give food unless he received pay for it. Esau had no money, but Jacob knew that one day his brother would inherit land from their father and decided it should be the price of a bowl of soup.

"Sell me this day thy birthright," he proposed.

Esau regarded his hardhearted brother pitifully, knowing that this was an unfair bargain. But the scent of the clay pot of porridge tantalized him and the pangs of hunger gnawed at his stomach.

"Behold, I am at the point of dying, and what good will this land do me unless I live?" he reasoned.

So, for a dish of stewed lentils and a piece of bread, Esau sold his inheritance.

Lentils were as common in those days as salt and flour are in modern households. So many were grown along the Nile River that one village was named Lentiltown. The little, hard, brown peas were used in trade to pay for purchases. An obelisk, one of the stone columns with which Egyptians commemorated important events, was transported to Emperor Caligula in Rome, and he sent back a boatload of lentils, 120,000 boxes of them, as partial payment.

The ancient Greeks thought lentils good enough for commoners but not quite fitting for a man in a high station of life. They wrote of one citizen, "Now that he is rich, he will no longer eat lentils. When he was poor, he ate what he could get."

Our name for this vegetable derives from the Latin *lentus,* meaning "slow," because it was believed that eating lentils produced heaviness of mind and made men deliberate and reserved, indolent and lazy.

The word "peas" comes from both Greek and Latin. The Greek verb *pisere* means "to pound or stamp," and *pisos* signifies "fine pebbles." Peas often were pounded before cooking, for they were the small, hard kind that are split and dried for soups or fed to stock.

When the Romans were introduced to peas they called them *pisum,* and the English later changed the name to peason and pease, from which we have our own word. However, the Romans preferred another of the Leguminosae cousins, the chick-pea.

Children of the Near East know it well. When they hear a peddler in the street with a basket of brown peas on his back calling, *"Leblebi, leblebi,"* they go with copper coins to bargain for a cornucopia of parched chick-peas to munch between meals, much as we eat popcorn.

Caravan travelers used to buy small sacks of *leblebi* to eat on long journeys in the desert, where only the most nourishing food could be carried.

In ancient times, when the Romans were out to conquer the world, their armies took chick-peas on marches, and thus this food reached Spain. Though the Romans were driven back, their chick-peas remained because they were good in stews. Then *garbanzos,* as the Spanish called them, went on new conquests. They were part of a ship's supplies when Columbus and his successors explored the New World. Wherever Spaniards settled in the Western hemisphere they planted chick-peas, and *garbanzos* became popular in all countries south of the United States.

Most of us probably would prefer chick-peas to the coarse field peas that were the only others known during the Middle Ages. Until 1700, no one except the richest families had seen the green peas we commonly eat. There was only the kind grocers call "split peas," and every European household kept a kettle of them simmering on the stove day after day. English children grew so tired of the stale porridge served in poor families that they poked fun at it in a song:

> Pease porridge hot, pease porridge cold,
> Pease porridge in the pot, nine days old.

Boiled field peas were tiresome fare, but on many occasions they saved the population of northern Europe from starvation. They were a principal part of the diet of armies, navies, and the merchant marine until the discovery of America, after which potatoes and navy beans were added to the eatables that could safely be stored for many months.

When Columbus landed in the West Indies, he changed the menus of the world. Foods unknown in Europe traveled back across the sea with him, and new plants were planted in the Americas. The first peas harvested in the New World were from seeds his men sowed at Isabela, Santo Domingo, in 1493. From this earliest European settlement in the New World, peas were carried far and wide. Soon Indian tribes of the north were growing them.

Then, in the year 1696, the royal courts of Europe, particularly the French, were buzzing with gossip about a new vegetable developed on their own continent. Somewhere in France a farmer must have discovered among his pea vines some with more delicate and better-flavored pods, and by taking pollen from these distinctive plants and placing it on blossoms of other vines, he started a new strain. The pods were filled with sweeter, greener morsels, which soon found their way to the king's table. Noblemen who dined with Louis XIV sent their servants to market seeking the new garden peas. What matter if they paid fifty crowns for a pint of pods? Had anyone ever tasted a more delightful vegetable?

The only way cooks knew how to prepare them was boiled in the pod and served unshelled. Flunkies brought platters of them to the royal table and befrilled and bejeweled courtiers set to work licking peas out of their shells.

Across the English Channel, ladies of fashion complained unless their husbands permitted them to send for the French "peacods" that were so expensive. Poets were enchanted by their flavor and wrote verses to "runcival pease," so called because they were slightly wrinkled.

Ever since field peas crossed the Channel during the Norman Conquest, Great Britain had been satisfied with old-style porridge. Now the English desired tender green peas. They saw that in the salty air of their islands the French vines developed wonderfully. This interested a botanist, Thomas Knight, and he experimented with them in 1787. By choosing pollen from one vine and placing it on flowers of a specially sturdy specimen, he created stronger and better plants. English growers who followed him produced other varieties, which were carried to the United States, and American breeders continued to improve them. A new kind had also reached America from another source; the black-eyed or cowpeas were brought by slave traders from Africa about 1675.

At first, American farmers wanted peas that would be good for both drying and eating fresh in the shell. They had not forgotten Europe's periodical seasons of hunger. Eventually mechanical inventions and the discovery that vegetables could be packed in cans or frozen eliminated the need for drying foods. Peas were now needed with pods that could be opened easily in the vining machine, a mechanism composed of reels and paddles operating in the same direction but at different rates of speed. The reels separated vines from pods, and the paddles knocked out the peas. The shells are usually stored in silos for stock feed.

The United States became the most important processor of peas in the world, and great fields were sown to supply the demands of canneries. Seedsmen next developed a variety suitable for freezing. Formerly, growers were most anxious that their peas have good canning qualities, particularly that they not burst during cooking. Freezing presented a fresh problem, for foods preserved in this way must hold their color and appear inviting when thawed. A special kind of pea that would fulfill this requirement had to be developed.

Peas intended for freezing must be harvested rapidly at the peak of their goodness. After passing through viners, bits of pods and other waste are winnowed out. Then they are washed, dunked briefly in hot water, and cooled. A belt carries them along an inspection and grading trough; they are drained, packed in cartons by automatic filling machines, and sent to the quick-freezing machine. From there, they go into cold storage, where they remain until shipped in refrigerated cars or trucks.

Some peas are frozen one at a time as each falls through a blast of cold air in a tunnel. These can be mixed with sauces and other vegetables before being packaged as complete dinner dishes, ready to heat and serve.

Man can now eat fresh green peas when the snow lies a foot deep outside his window. If Esau had lived in our time, he would have demanded something much better than a miserable clay bowl of pottage in exchange for his birthright.

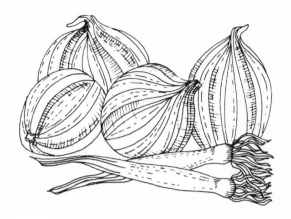

3.

The Onion Family

Early in the world's history, a band of wanderers reached the northern shore of the Red Sea and entered the Nile Valley. At that time the delta in this part of Egypt was not so large, and the river filled the valley. On its banks a few grains and vegetables grew wild, but there was not enough grass for the colonists to lead the pastoral life to which they were accustomed.

They were forced to get acquainted with new and strange foods. They gathered leaves of a hedge plant and the roots and seeds of the lotus, or water lily, which grew in great abundance in the marshes. Sweet edible bulbs of aquatic plants were their dessert. They made bread from dried lily seeds crushed into powder. Children, fed roots and seeds of rushes and lilies from infancy, sometimes grew up without ever having eaten other substances.

One of the edible plants of the valley was the papyrus, or paper rush. When the best portions had been selected for making mats, paper, and fibers, the pith from the lower part was set aside to be boiled or roasted. The roots were eaten raw.

Until traders arrived from eastern and northern lands, the first inhabitants of Egypt lived on this simple fare, combined with a little fish and game. Gradually, merchants from Assyria and Asia Minor found their way to the Nile, bringing cereals and vegetables such as peas, radishes, broad beans, cucumbers, and various kinds of onions.

The valley had rich earth for farming and the new seeds grew well in the regions that were flooded every year. By means of canals the Egyptians carried water away from the river and widened the belt of garden land. The Nile rose and fell with the seasons. As soon as the rains were over, the water went down, leaving great mud flats to be cultivated. In November, when these had dried enough to bear a man's weight, farmers sowed seeds broadcast, raking mud or dragging branches to cover them. The harvest was in the spring. Water in the canals was lowest in early June. In July it began to rise and continued to rise until the middle of October, when it receded again.

Caring for vegetables must have been one of the first things Egyptian farmers learned to do. As they became skilled they planted onions, leeks, garlic, and cucumbers in long strips on the banks of canals and around reservoirs. They also had fields of peas, beans, and lentils, and grew lettuce and radishes for the oil that could be pressed from the seeds.

While the men worked in the fields, the women prepared the produce for market, and brought it to cities in baskets balanced on their heads or in bundles carried on their shoulders.

When the floods occurred in summer, water covered the fields and the villages stood on islands. For three months the vast army of farm laborers had nothing to do. Their kings, the Pharaohs, made use of this free time of the people by putting their subjects to work building pyramids, great monuments to serve as royal tombs. At the beginning of the dry season farmers would leave home and go to the raised roads along which huge stones were being moved to construct the pyramids. No great individual skill was needed to push these stones along on rollers. Several men could handle each block. So the farm laborer simply packed as much food as he could carry and went off to eat it on a moving job instead of consuming his produce at home. When he ran out of supplies, someone would bring him more.

It is said that onions, radishes, and garlic built the pyramids. The people of the Nile ate quantities of onions because they grew abundantly and were cheap. The ancients were sure that a vegetable that had the power to make men weep must possess strength-giving properties, and that therefore it was fitting food for workers engaged in building gigantic structures.

When an Egyptian died, onions often were placed in the hands of the mummy so that he could enjoy his favorite vegetable in the next world.

Although the poor took onions to temples and offered them to their gods, the priests would not eat these vegetables. They were even more superstitious about garlic because of its strong flavor, calling it unclean and refusing to allow pictures of it to be inscribed on the pyramids.

Three kinds of onions were grown by the Egyptians. All must have been brought from the high plateau of Turkestan near the Caspian Sea. To the east of that region lay the Onion Mountain chain, so named because the plants grew wild there.

After the Israelite tribes had been driven out of Egypt and were wandering in the desert, their mouths watered for Nile Valley onions, garlic, and leeks, which they had eaten with barley bread. After many years in the wilderness they reached the land of Canaan, and were overjoyed to see these vegetables again, growing in abundance.

An onion hitherto unknown to them also grew in the Holy Land, the small green shallot, or scallion, which took its name from the city of Ascalon. Some botanists think scallions grew wild near Ascalon only by accident, and that a clever gardener developed them from the common onion.

There is a fifth variety, the mild little chive, with purple flowers, grown in herb gardens for its tasty tops. It is native to many parts of the Northern hemisphere.

All of these belong to the lily family, but the strong scent of most of them gave rise to curious ideas and superstitions not generally associated with such mild-looking plants. When Alexander the Great saw onions in Egypt, he ordered the Greeks to cultivate them so that his soldiers could eat them and thereby gain strength and courage. The Romans were not sure that anything less than garlic would have this effect on human beings, so they placed their faith in the pungent little buttons and made sure that all laborers and warriors consumed quantities of them.

The feeling of the Greeks toward garlic was somewhat different. They reserved it for criminals, thinking it would purify them. But a Greek gentleman would not hesitate to send a handsome jar of onions to his best friend for a wedding present.

In ancient Greek the word "onion" also signified the head of a man. It happened that a group of colonists sailed to Italy, pretending they were on a friendly mission. The natives did not trust them, but the Greeks swore solemnly, "So long as we still tread earth and wear heads upon our shoulders, we will share the land with you in peace and friendship."

The Italians trusted them, laid down their arms, and welcomed the strangers.

The Greeks, however, had come to conquer, and as soon as the inhabitants put aside their spears and returned to the fields, the invaders, according to legend, removed their sandals and emptied dust from them. Then they pulled from their shoulders bunches of onion heads that had been hidden beneath their tunics. By these acts they considered themselves released from their oath and, attacking the natives, drove them out of that part of the peninsula.

Garlic, leeks, and onions were used in medicine from ancient times. Their penetrating taste and smell gave rise to the belief that they would counteract poison or break a charm. In the Middle Ages onions often were hung in the center of a room to chase troublesome maladies from the house. There were other curious ideas: the Emperor Nero, for instance, ate leeks to clear his voice.

The leek is the national emblem, or badge, of Wales. It commemorates a sixth-century battle in which the victorious Welsh wore leeks in their caps to distinguish themselves from the enemy.

Old books on fortune-telling advised a maiden anxious to know who her future husband would be to set several onions near a chimney and name each for a suitor. The bulb sprouting first would indicate the lucky man.

The people of almost every nation in the world eat onions. They are excellent as flavorings, and food-processing plants in the United States use quantities in the manufacture of soups, sausages, sauces, and the many relishes and pickled preserves Americans consume.

Onions reach our tables in some form every day, lending zest to salads, meats, and dressings. In spite of their strong odor and tear-provoking power, they are one of our best-loved foods.

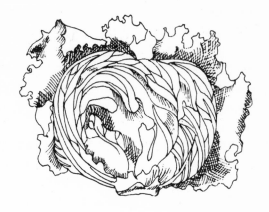

4.

Lettuce and Cress

Several hundred years before the time of Christ, Alexander the Great, a young Greek king with an ambition to rule the world, led his army into Persia on an expedition that was to have far-reaching consequences. Galloping day and night, his men pursued the great warrior Darius III toward the Caspian Sea, only to find on their arrival that the Persian was already dead, killed by Bessus, one of his own generals who planned to rule in his stead.

Alexander had no intention of seeing anyone but himself reign over the Persian Empire, so he prepared to follow Bessus into the latter's own country in the Oxus (now called Amu Darya) Valley. All that winter the Greek rested his troops by the Caspian Sea and in the spring set out for Afghanistan. Before he was across the

high Hindu Kush Mountains, storms blocked the passes
and he was once again delayed.

By that time his men were unhappy. Accustomed to
sunny skies and a mild climate, they were terrified by the
long, gloomy days amid blizzards in the dismal wild
mountains. They eagerly awaited the days when the ice
would melt and they might pick their way through the
canyons and descend to the Oxus. When they were once
more on the move, they found Bessus had ravaged the
country and left nothing for them to eat. Footsore and
hungry, they marched on.

Suddenly they found themselves in a wide valley unlike any other place in the world—a natural garden so
fertile that many believed it must be the earthly paradise.
Broad beans, lentils, peas, chick-peas, radishes, onions,
cabbage, turnips, spinach, carrots, beets, and various
grains grew wild in flowering meadows. Far ahead
stretched mile upon mile of irrigated plantations abound-
ing in melons, figs, almonds, cucumbers, squashes, and
eggplant. The inhabitants of this region were the most
fortunate of the early tribes on earth, for this wild garden supplied all the nourishing food they needed. The
land seemed to invite men to become agriculturists, and
farming was their most honorable occupation.

The Greek warriors were delighted with the fruitful
country, and, after capturing Bessus, Alexander permitted
a number of his men to remain and administer the new
domain. He founded the kingdom of Bactria on what is
now the northern frontier of Afghanistan and chose as
capital Balkh, a place so old that it was called the
Mother of Cities. Here he married an Asiatic princess

and lived for a short time in the splendor of an Oriental potentate. Reluctantly he at last left the delights of the Oxus Valley and took the perilous road over the Hindu Kush to continue his conquests and seek the fabled riches of India.

He had enjoyed the good things in the valley far too much to want to do without them in the future. Now he continued to devote many hours to the pleasures of dining. While he sat in a silken tent and devoured dainty foods served on golden platters, squads of his men explored the surroundings. They made maps and listed animals and vegetation, particularly the kinds that might be suitable for transplanting to Greece.

Wherever Alexander went he indulged his enormous appetite for food and knowledge. After tiring himself with thirty- and forty-day marches, he made camp in places where food was abundant and commanded his officers to join him in great banquets that lasted for days. While he was wining and dining, his busy scribes worked on, jotting down the uses of rice and lemons, bamboo, cotton, and myrrh, and the latest dainties on his dinner trays.

In this manner the Greeks made a marvelous collection of data that served as the foundation upon which the ancient world later built its commerce.

Although Alexander died of malaria on his way home, the paths he followed on his conquests became the caravan roads between East and West. Many of the edible plants of the Oxus Valley reached the Mediterranean over these routes.

One of the things that most impressed Alexander was

the way eastern nations used salad plants. Seventy years earlier the Greek soldier-historian Xenophon, who was in the service of the Persian prince Cyrus the Younger, ordered his soldiers to eat garden cress when they were on the march. He claimed it had unusual properties and that Persian mothers were advised to feed it to their children so that they would become tall, straight warriors.

In ancient times cress grew in widely separated places. It is thought to have originated in southwestern Asia, probably in Iran, and to have spread from there. A favorite variety is watercress, which grows in running water, and is found in all temperate regions of the north. It was adopted by the Greeks and Romans as a salad but was not cultivated to any extent until the sixteenth century. Today the English are fond of it, eating the leaves for breakfast or enjoying cress sandwiches with their tea. Watercress is also a favorite of Americans and is found in stores in spring and summer.

Although the ancient races of the East had never heard of the vitamins stored in salad plants, they liked green foods. People often plucked and washed the leaves and devoured them on the spot.

There is a Persian proverb, "Eat cress and learn more wit." Pliny the Elder, a Roman naturalist, had another idea about the properties of cress. He said, "Eat vinegar on cress and you will never be insane."

The early kings of Persia thought lettuce should be served at the conclusion of a meal. After the meat, grain, and vegetable courses were cleared away, large platters of fruit were brought in. Among the rosy apples, plump melons, grapes, and peaches nestled clusters of romaine. Sometimes there were other garnishings, for many kinds of chicory, cress, and endive grew wild in that part of the world.

Such a tray of fruits and salads once cost a Persian princess her life. She was the daughter of Cyrus the Great. Her brother Cambyses was then king of Persia, but she had another brother, Smerdis, of whom she was very fond.

It was the custom among Oriental rulers to put out of the way all relatives who might have their eyes on the throne; therefore, when Cambyses succeeded his father, he ordered that Smerdis be executed.

A few days later Cambyses sat at dinner with his sister. When the meal was cleared away and the fruit was brought on, the princess chose from the tray a beautifully formed head of lettuce. Taking it up deliberately, she stripped away the leaves one at a time and then asked her brother what he thought of it.

Cambyses replied that the lettuce did not look good without leaves.

His sister said instantly, "It is that way with our family since you have cut off a precious shoot."

Cambyses was angry. He knew that the princess was accusing him publicly of having murdered Smerdis, so he had her imprisoned and condemned her to death.

Cultivation of lettuce spread from the Greek Empire to all the countries in southern Europe and northern Africa where the climate was suitable. One Greek scholar gave his neighbors something to talk about when he went to his garden daily and sprinkled his lettuce with wine, which he thought would improve its taste.

The Greeks and Romans associated many curious beliefs with their vegetables. They said that endive would cure headaches, and their physicians prescribed lettuce for wakefulness. The Romans were supposed to have grown a kind of black lettuce that made people sleepy. Lettuce is known to calm the nerves; the juice contains a sedative that reduces tension. Lettuce is only 5 or 6 percent organic material and the rest is water. Greeks and Romans squeezed out the milky juice and drank it as a cure for insomnia. The liquid was called *lactuca* and from it lettuce got its name.

Once, when the Emperor Augustus Caesar was ill with a dangerous disease, his physician cured him with lettuce. The vegetable contains vitamin A, which helps resist infection. It also contains vitamins B, E, and B_2, as well as calcium and iron. After the emperor's experience, the Romans considered this plant very valuable and attempted to grow it in all seasons so they would never be without it. They used it in both pottage and salad.

The Greeks found a way to make lettuce grow in tighter, paler heads by heaping sand around the plants. Now that vegetables have been greatly improved by modern experts, the lettuce we eat is much more tailor-made in its appearance than in the past.

There are several types of lettuce, the commonest being the crisphead, or iceberg, and the long-leaved romaine, also named Cos from the Aegean island which in ancient times was famous for growing it. Then there are loose-leaved kinds, the butterhead and a stem variety called celtuce, common in China.

Although lettuce of some kind was introduced into

England at an earlier date, it was not cultivated much until Queen Elizabeth's reign, when it was respected as a food "to cool the hot stomach." This delightful salad is a close cousin of endive and also of the dandelion, which is used in some communities for its tasty and nourishing tops. All belong to the chicory family.

The leaf called chicory, or succory, has for ages served as food for man and beast. In Belgium its tender young roots are regarded in the same way as carrots. Dried chicory roots have been employed in medicine. The young leaves are used for salad, as a cooked green, or as an adulterant for coffee. Chicory also makes excellent fodder.

Lettuce was introduced into America soon after the first colonies were established. It is now our most important salad plant and one of our largest vegetable crops. It is grown in vast fields and is harvested by a packing machine which trundles along the rows. Men follow it, cutting off ripe heads, and women riding the harvester lop off outer leaves and swath the firm portion in transparent plastic wrappers. Men standing in the center of the machine can pack two hundred cartons an hour, leaving them for trucks to pick up and carry to distribution points. More than 85 percent of the many millions of dollars' worth of lettuce the country produces every year comes from the Rocky Mountain region and the Pacific Coast.

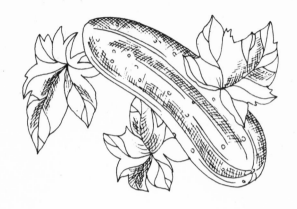

5.

Cucumbers

Among the plants that Alexander's soldiers saw growing in the Oxus Valley and in Persia were edible gourds such as cucumbers and squashes. To ancient man these oddly shaped fruits were puzzling. Were they herbs or vegetables or mysterious children of the earth itself? A large, round gourd somehow suggested a baby planet that had just been hatched. In those days both squashes and cucumbers grew to enormous size, and an early writer likened the latter to serpents coiling in the grass and swelling to "belly shape." Stories that have been handed down about their former size and shape are not unbelievable, for as late as the Middle Ages these fruits were larger, rougher, and less symmetrically formed than they are now.

Cucumbers first sprang up in northern Africa or in Hindustan, but they proved so useful and popular that they spread to western Asia about three thousand years ago. They were among the vegetables the Israelites missed after being driven out of Egypt. Later, cucumbers became a favorite food of Greeks and Romans. Their present name derives from the Latin *cucumis*. It is said that the Roman Emperor Tiberius kept his gardeners busy forcing cucumbers in beds so that he might have them on his table every day in the year.

Although many members of the gourd family are native to America, the Western hemisphere had no cucumber. The vine we call "wild cucumber" belongs to another family. Even the small, prickly bur gherkin found growing wild in the West Indies is thought to have been imported from Africa by Negro slaves in colonial days.

Columbus raised the first cucumbers in the New World in Haiti in 1494. Other colonists carried them to North America. The French took them to Montreal, and the British took them to Virginia when the initial settlements were made there.

Besides eating them fresh in salads, Americans consume cucumbers of every size and shape in the guise of pickles. Their use in Europe did not spread as rapidly as one might suppose, and as late as the fourteenth century cucumbers were still a curiosity on the northern part of the continent. Herb doctors in the Middle Ages recommended them as a cure for red and shining noses, pimples, and similar skin ailments, advising women for the sake of their complexions to eat cucumbers thrice daily, baked in pies or boiled with mutton or mixed with oatmeal porridge. At the end of three weeks the blemishes were supposed to have disappeared.

For many centuries cucumbers were most highly prized by the inhabitants of southwestern Asia. They ate them raw, boiled, baked, or combined with other foods. A Persian or a Turkoman might start his breakfast with them. Tartars, Slavs, Mongols, and Russians—all those who invaded the Oxus Valley—adopted them as a favorite dish.

Cucumbers like sunshine and water, and these were not lacking in the valley. Although rainfall was not heavy, the inhabitants found a way to provide an abundance of water. In the foothills of the mountains they built barriers to hold back the snow so that it would melt more slowly and drain into canals. They turned streams into new channels and conducted the water across plains. Throughout southern Turkestan and northern Afghanistan and Persia, where farmers lived at a distance from the mountains, they made special efforts to bring water to their dry lands. Usually beginning near the summit of a mountain, they dug a series of storage pits or reservoirs and joined these by means of underground aqueducts, called *kanats*. They were covered over so that the hot, dry climate would not cause too much of the melted snow to evaporate. This system of irrigation is still used in Iran.

When the water reached the thirsty farmland it was lifted from irrigation ditches by means of skin containers attached to ropes. A man would drop a pouch into the canal and signal his oxen to pull. As the animals moved forward, the heavy vessel was raised and the man emptied the water into tiny channels that led off through the fields.

Inhabitants of the Oxus Valley did not know what a

bucket was until they saw an Oriental traveler using one. If the farmers did not have buckets, however, they were ingenious enough to build waterwheels where the current of the Oxus and its tributaries permitted. The wheels emptied scoopfuls of water into troughs that conducted the precious liquid to the fields.

Irrigation was considered noble work, and kings gave it their full support. The religion of the people counseled them to take water from where there was too much and put it where there was too little. So in the entire region, river water was spread over the plains by a clever system of canals, among the earliest constructed by man. As a result, the plain between the Oxus (Amu Darya) and the Syr Darya Rivers, in today's Uzbek Soviet Republic, became so rich and green that in ancient times there was no other land like it. Fruits and vegetables reached unbelievable size, and the largest melons in the world came from there.

In the midst of this rich area lay the city of Samarkand, which the Holy Book of the Persians called one of the "places of abundance." Its name meant "Fat Market" because so many foods grew around it. Outside the walls in every direction were miles of gardens consisting of rows of fruit trees and vegetable patches, with flowers blooming among the edible plants. There the yellow cucumber and melon blossoms spread out amid rosebushes. In every part of this garden land running water rippled through small canals or was sprayed with a tinkling sound into square basins or round ponds. By the river it splashed out of waterwheel scoops and ran murmuring through troughs.

It is difficult to realize that, without irrigation, parts of our earth would quickly become desert again, yet many of the fruit, vegetable, and flower gardens that Alexander's soldiers saw are now dry and covered with weeds. If we were to prod in the sandy soil, we would find traces of aqueduct walls and canals. In the thirteenth century the land between the two rivers, which had been farmed by one of the earth's oldest races, met with a calamity from which it never recovered. At that time Genghis Khan and his hordes rode out of the east and destroyed everything in their path. The monster melons, crisp cucumbers, delicious eggplant, and fields of peas meant nothing to the Mongols except food which the enemy should not be allowed to have. If the conquered inhabitants were hungry and obliged to turn their attention to mending their farms, they could not fight back.

So Genghis Khan chose the quickest way to ruin his victims: he wrecked their irrigation ditches. The land was never the same again. Now much of the country is an open plain with scattered dwellings, and it would be difficult to believe that gardens had existed there were it not for the blossoms that spring up each year. If you inspected them, you would find not all are weeds; many are vegetable plants growing wild—peas, turnips, cress, and others.

As for the cucumbers, there is still no village without them. Even the conquering Mongols had to admit that these were indeed tasty, and that in the future they must have this vegetable in their own country.

Throughout the Far East, cucumbers are now a favorite food, enjoyed by all ages and usually preferred unpeeled. Those not eaten fresh are used mostly in soups. The Chinese force them out of season and go to great lengths to make them grow straight, having been known to hang stones by strings to the ends of their cucumbers so that they will turn out well formed.

Egyptians used to make cucumber water. A hole was cut in a ripe cucumber at its base. After stirring the insides with a stick, the hole was closed and the cucumber buried in a little pit for several days "after which the pulp will be found converted into an agreeable liquid." Cucumber water was apparently drunk for pleasure.

Cucumber juice has also been used as an ingredient in cosmetics.

6.

Radishes

Perhaps the radish can claim as its native home more countries than any other vegetable, for in earliest times it was cultivated throughout Asia, from China to Syria, and on down into Egypt, where the pyramid builders ate radishes as part of their daily ration.

Thousands of years ago the radish was not the crunchy appetizer we delight in eating at the beginning of a meal. It was grown for the nutritive oil in its seeds, and in the Orient today one variety still furnishes this product.

In ancient times radishes grew exceedingly large. A Jewish legend tells of one so huge that a fox hollowed it out to live in. The Greeks and Romans told stories of radishes weighing from forty to one hundred pounds. In those days farmers removed the leaves so that the size of the roots would increase. When ripe, the great radishes were roasted in ashes or cut into soup or fed to livestock in winter.

As the centuries passed, tastes changed. In the Mediterranean basin the desire was for smaller radishes, whether long or round, white, red, or black, as the Spanish ones are. Giant radishes disappeared from the West and tales about how large they had once grown were considered fables. Then, in the fifteenth century, Vasco da Gama discovered the sea route to India and gradually ships pushed along the Asiatic shores until they reached China.

The Chinese were secretive and suspicious. They wanted no strangers intruding upon their national privacy. If Western sea captains wished to come to their shores, they must anchor at Canton and receive in trade only those products the Chinese chose to send.

For many years everything about China was a complete mystery. How did the people live? What did they eat and drink? No one could say. Tea and ginger were food products they exported, but the outside world knew little else about the relation of agriculture to the daily existence of the Chinese people.

Although China seemed eager to receive products from other lands, its own were carefully guarded. Since earliest times it had been a country of too many inhabitants and too little fertile earth. Shut off from the rest of the world by seas, mountains, and deserts, it lacked a variety of food plants. Millet, panicum (a grain), soybeans, rice, certain roots, and perhaps pe-tsai, or Chinese cabbage, were the main crops. Legend says that many of these were the gift of Shen Nung, a great emperor called the Divine Cultivator.

Nearly five thousand years ago, according to the story, Shen Nung called his followers together at the foot of the Mountain of Cereals and showed them how to sharpen a stick and drag it through the ground to plow a furrow. He taught them to plant five grains and one hundred useful grasses, among which were medicinal herbs. From then on, until 1911, when the reign of the Manchu emperors ended, it was the custom to honor Shen Nung each year at a ceremony where rice, wheat, and two kinds of millet and soybeans were sown.

This served to remind the Chinese of the importance of agriculture to their nation. Theirs was a vegetable civilization. Mines, timber, and herds played only a very minor part in Chinese life. Having no forests and no game, the people were obliged to grow their food; farming was not a chosen occupation but a necessity. There were not enough greens for both man and beast, so dairy products were lacking. Peasants ate whatever the land afforded, even if it was only a water plant, leaves of alfalfa, or tender shoots and buds of shade trees. Fish, birds, and pigs were the only things that gave variety to their food.

Each tiny farm was forced to produce the maximum quantity of vegetables. To us many of them seem tasteless.

Botanists knew that these farmers must hold the secrets of many fruits and edible plants, but for centuries it was impossible to secure samples. Sea captains brought back a few from Canton and Macao; others were smuggled overland in Russian trading caravans and carried across Siberia and the Urals to St. Petersburg.

Probably an East India Company ship brought the first Chinese radish seeds around the Cape of Good Hope to the keeper of the "physic garden" of London's Apothecaries Company. There at the opening of the eighteenth century the radish must have been grown as a medicinal plant.

A few years later Linnaeus, the Swedish botanist to whom we owe most of the plant classifications used in the scientific world, secured employment as director of Hartecamp in the Netherlands. He was then a young man, impressed with the responsibility entrusted to him. Hartecamp was a most important center of plant culture. Therefore the keeper made detailed notes of his work and endeavored to secure all kinds of foreign seeds. His notions of geography were vague, but eagerness to cultivate strange plants outweighed this shortcoming. By persistent effort he procured ten kinds of Chinese seeds, some brought by a Swedish sea captain. Linnaeus must have been astounded at the enormous roots that grew from the insignificant seeds of Lew Chew radishes.

After the celebrated botanist left Hartecamp and returned to Sweden, his interest in Oriental flora continued. His son was devoted to the same pursuit and between them they identified 160 Chinese plants.

Every traveler to the Orient in later years spoke with amazement of the giant radishes, and when Commodore Matthew C. Perry paid his history-making visits to Japan in 1853–54, he too was astonished to see them growing two and three feet long and a foot in circumference. The Japanese salted and pickled these twenty-pound vegetables. In China radishes generally are eaten cooked.

Radishes, which belong to the mustard family, are low in calories, and long were considered more suitable for home gardens than for market produce. Henry VIII liked them, but in his day they appeared on the table roasted or boiled, or in sauces for meats.

Among the Greeks the radish was greatly respected. Radish syrup was used as medicine for whooping cough and hoarseness, and persons who slept too much were advised to chew radishes.

It was the custom to bring likenesses of foods wrought in precious metals as offerings to the temple of Apollo at Delphi. Models of beets were made of silver and turnips were molded in lead, but the radish was made of gold, which shows how much it was admired at the time.

In Java and India the natives use a large root called the serpent radish.

In the United States we have the horseradish, a taller and coarser relative of the common radish, whose fleshy root is used only as a condiment because of its pungent flavor. It was brought here from Europe. Growing wild from Finland to Astrakhan, it was carried to England with the Saxon invasion, long before the Norman Conquest.

7.

Spinach

One of the teachings of Shen Nung, China's Divine Cultivator, was that his countrymen always should accept gratefully the useful plants other lands had to offer. Eventually, by following this precept, the Chinese became great farmers and masters of plant economy, with three quarters of the population engaged in agriculture.

Their first contact with other civilized nations was an adventure in the second century before Christ, which opened the road by which spinach and many other vegetable foods traveled eastward.

Between China and the peoples of the West stretched high mountains and broad deserts. The Chinese were a stay-at-home race, busily trying to feed an ever-growing population. They produced silk, lacquers, and porcelains, but were completely ignorant of many of the commonest plants and animals of the ancient world.

On the fringe of their empire restless nomad tribes, driven by the need for richer grazing lands, disturbed the peace of the Chinese by frequent raids. One day some enemy tribesmen taken in battle told the Emperor Wu Ti of Han a story that was to change Chinese civilization. They said that their Tartar chief had defeated the king of the neighboring Indo-Scythians, killed him, and made a drinking cup of his skull. The Scythians had fled far to the west of China and were scheming for revenge, but they had no ally to join them and they were not powerful enough to attack the Tartars alone.

This was good news to Wu Ti, for he had fought the Tartars alone in the past and believed that if he could get a message to the Scythians, they would help him to wipe out the enemy. He decided to send an ambassador to them and chose for this errand an officer of the imperial household, Chang Kien.

Accompanied by a lone attendant, Chang Kien set out in 138 B.C. to find the Scythians, but on his way west he had to cross Tartar territory. Within a few days the enemy captured him. Chang Kien conducted himself agreeably, so he was well treated, but he was not permitted to leave. Months dragged into years and the Tartars would not release him. They married him to a Tartar woman, by whom he had a son.

Although Chang Kien seemed content to remain, he still kept hidden in his clothing the emperor's token showing that he was an ambassador, and he waited for his chance to get away. Ten years passed and the Tartars were sure that the Chinese had forgotten his imperial mission and would remain among them. When they no longer watched him closely, Chang Kien escaped.

Instead of turning back he pushed westward and arrived in a new land where the people had heard of the wealth and fertility of China and had long wished to communicate with its emperor.

"If your chief will send someone to conduct me to the country of the Indo-Scythians," Chang Kien told them, "and if I succeed in reaching it, on my return to China my king will reward you with untold treasures."

The strangers gave him an escort, and Chang Kien thought with their help he would soon reach the end of his journey. He did not know that during the ten years of his captivity the Scythians had moved many times.

The little party started south and then west. Hundreds of miles dropped behind them and the rivers flowed in another direction. Yet the Scythians were still far away toward the sunset. It was not until Chang Kien crossed the Pamir plateau and descended through the Hindu Kush Mountains that he found the people he was seeking. The Scythians had raided the walled cities and irrigated farms between Samarkand and the Oxus River, driving the Greeks out and putting an end to the kingdom of Bactria, which Alexander had founded two hundred years earlier. They told Chang Kien they did not desire to go back over the long road to China and again take up arms. They were willing to forget that a Tartar khan had made a drinking cup from the skull of their king. By forcing them to flee, this khan had done them a favor, for the Scythians' old lands had not provided ample food. Now that their tribes ruled the most fertile spot in central Asia, they intended to settle down and live in peace.

For a year Chang Kien remained among them, urging that they go back and help Wu Ti with his wars, but it was of no use. He saw that he must return home without allies for the Chinese. Nor were his dangers over, for he must pass once more through Tartar lands. He said good-bye to the Scythians and began his long journey over mountain, steppe, and desert. Once again he fell into the hands of his enemies. The Tartars took him to their camp and again treated him well, but would not let him go. Only a few days more and he would have been in China. Now would he ever reach there?

Another year passed, and then the Tartar khan died and a prince stole his throne. The Tartars quarreled among themselves, and while they fought over who should be khan, Chang Kien escaped and at last reached the court of Wu Ti.

By this time he was wondering how the emperor would receive him. He brought no promises of allies, but in the folds of his robe and engraved in his memory he carried precious gifts which he hoped Wu Ti would see fit to accept. The trophies he brought were seeds of the grape vine and alfalfa, and the recollections he carried in his mind were soon set down in a report, "Record of Remarkable Things in Foreign Countries." Among them were descriptions of a cloth woven of cotton and of a useful cane called bamboo, both of which had reached the Oxus from the unknown land of India.

Not since the age of the Divine Cultivator had a Chinese emperor been given the chance to make great agricultural gifts to his people. Wu Ti saw that Chang Kien's service was worth more than a dozen armies. He appointed the traveler imperial chamberlain and sent him again to bring more knowledge of products of the West.

Chang Kien was a true explorer. Willingly he braved hardships of the trail, but this time he had a caravan of one hundred men to carry goods and a famous huntsman to provide meat with his bow. Chang Kien went to eastern Iran and gathered information about lands beyond to the west and south. Thirteen years later he returned to China and told Wu Ti he would like to be sent to India, where there would be rare produce to exchange. The emperor promised instead to organize more parties to explore the trade routes; he needed Chang Kien as his minister of state. The brave explorer was given the title of Great Traveler, but by then he was old and did not live to see China reap the benefits of his journeys.

Chang Kien became a hero to his people, and as exploring parties brought back more things from the lands he had discovered, the Chinese credited him with giving them almost every useful vegetable, fruit, and nut that was new to their country. Although Chang Kien himself did not carry many plants to China, he had much to do with their introduction. Because of his suggestion that trade routes be sought, the emperor every year sent from six to ten caravans with gifts for rulers of the western lands which Chang Kien had told about. Some of these expeditions were nine years on the road, and when they returned they often brought seeds and plants among other presents.

Many years were yet to pass before merchant caravans made regular trips across Asia. The few that went each year were too busy protecting their cargoes from robbers and enemy tribes to give attention to farm produce, which was more perishable than other goods. They carried only articles of great value that could be packed in a small space.

Meanwhile the Chinese went on eating rice, fish, soybeans, lily bulbs, and long, loose cabbage. They had onions, the rape turnip of Mongolia, and, from some mysterious source, the sweet potato. To these were gradually added the foods that came over the new trade route from the West—cucumbers, peas, broad beans. Very slowly the diet of China was enriched by new plants, but it was not until the seventeenth century that people of the East actually became acquainted with the majority of our common garden vegetables. By that time the empire spread over Chinese Turkestan and other conquered lands and numbered among its allies the kings of Tibet and Nepal. The emperor, wishing to bring all India to his feet, made plans to invade those kingdoms which had refused to pay homage to him.

At this time he issued a peculiar request, demonstrating that not even during war could a Chinese ruler forget the teachings of the Divine Cultivator. He invited each tributary nation to prove its loyalty by giving China samples of the choicest vegetable products in the kingdom.

One ruler sent the wine vegetable, so named because men likened its taste to good wine. The plant was romaine lettuce. Another presented beets; another gave scallions; and the king of Nepal dispatched a dark-leaved plant known as *asfinaj,* or *aspanaj,* which he had received from Iran and held in high esteem. This herb, he had been told, was both food and medicine. Its usefulness already was known in the West, for Arab travelers had carried it from Persia and spread it to the Mediterranean Sea. They called it *sepanach.*

Thus both East and West became acquainted with the Persian herb, and in the fourteenth century Buddhist monks were eating it in China while Catholic monks were planting it in monastery gardens in Spain, to be used on fast days. In another hundred years it would spread to northern Europe. The English learned about it from the Dutch, who recommended it stewed or made into tarts or used to flavor meats and salads. People liked it because it was inexpensive and nourishing. They did not understand, as we do today, that it contains iron needed for building red blood cells, roughage for stimulating passage of food in the digestive tract, and three important vitamins.

The English called it the "Spanish vegetable," but as the years passed they borrowed that nation's word for it, *espinaca,* and changed it to "spinach." It has also been thought that its English name originated in "Spanacha," as coming from Spain. Whichever may be the case, we know that the name originally came from the Persian herb that was a king's gift to a Chinese emperor.

When Captain James Cook charted the South Pacific and discovered Australia, he found an equally useful plant, New Zealand spinach. It is not closely related to true spinach, but is used in the same way. Later it was found wild in Japan, Australia, and South America. Now it grows in gardens in the United States.

8.

Beets

Early travelers to Persia noticed that the people sweetened their drinks not with the sugarcane of India but with manna. This was not the kind the Bible tells us the Israelites found in the desert. Biblical manna may have been an edible fungus; the Persian variety was a juice that oozed from the camel's-thorn and dwarf oak bushes at the end of summer. It was sometimes called "sweet dew," and no Persian kitchen could get along without it.

Today it seems odd that the country was obliged to use manna for sweetening when it was the native home of the sugar beet.

Not knowing that all around them grew a vegetable that is one of the most important sources of sugar, the people waited patiently for the season of manna. With the first hint of autumn weather, this juice appeared suddenly on the plants at night. It had to be gathered quickly in the early morning, for it was quite perishable.

Manna was used like honey, and much of it was made into syrups flavored with fresh fruit juices.

In that day wild beets were considered useful only for their greens. Gradually, other countries began to cultivate them, and three or four centuries before the time of Christ they reached gardens in all the lands between the Mediterranean and Caspian Seas and as far away as western India. Some nations liked both root and leaves, finding them good for medicine as well as food. Their juice was supposed to cure headaches.

Later, Arabs traveling to the Orient said the Chinese could find no name for the plant when it was introduced there, so they simply called it "sweet" because of the taste.

Our name for the vegetable comes from its seed cluster. When this swelled and ripened, the ancient Greeks saw its resemblance to the letter beta in their alphabet, so that is what they called it.

Although the beet was praised for its sweet taste, no one suggested that sugar could be extracted from it. Then in 1747, Andreas Marggraf, a Prussian chemist, announced that the vegetable contained sucrose, the same delectable substance as in sugarcane. Marggraf did not carry his experiments far, but his pupil Franz Achard continued them and found out how to extract sugar from beets. He became so excited he foolishly declared the vegetable had other marvelous properties and could produce tobacco, molasses, rum, coffee, vinegar, and beer.

The Institute of Paris heard about Marggraf and

Achard and appointed a committee to find out if there was any truth to the statements. The committee was scornful. After its members made a few experiments, they said Achard's claims were ridiculous. His theory was discredited, and the possibility that sugar could be obtained from beets might have been forgotten had it not been for Napoleon Bonaparte.

In 1809, during his wars against Great Britain, the emperor decreed that France was no longer to purchase any materials from the West Indies. The French needed cane sugar from the islands. Napoleon said they could obtain sweetening from other products. He started factories to make syrup from raisins and honey. These were satisfactory for cooking purposes, but there was no crystallized sugar for the table.

Napoleon was forced to listen to so many complaints that he realized something must be done at once, because the stomachs of the French controlled their tempers. If he could give them the sugar they desired, he might be able to keep them contented. He appointed a board to find some way to procure sugar without engaging in foreign commerce. One of the members was a man named Nicolas Deyeux, chief pharmacist to Napoleon. He had heard of Achard's experiments and wondered if he could be right about sugar in beets.

Deyeux conducted fresh experiments and succeeded where the Institute of Paris had failed. Napoleon immediately established an imperial sugar-beet factory, and in 1812 the industry was under way.

There are white beets and red beets. In the eighteenth century Silesian farmers raised the pale kind as a forage crop for livestock. The stalks, leaves, and midrib of the leaves of one variety, Swiss chard, which does not develop a thickened root, are enjoyed in Europe and America.

When the manufacture of sugar from beets was seriously begun, efforts were made to increase the sucrose in the roots. By selecting seeds from various crops over a long period, experts were able to breed in the white beet of Silesia four times as much sugar as it originally contained. This specially developed beet may contain as much as 20 percent sugar. It is such an excellent source that almost as much sugar is made from beets as from cane. Much of Europe depends on its own beet fields for this product, and a large quantity is produced in the United States. Sugar beets have become one of the world's important crops.

Meanwhile their red relatives reach our tables fresh, canned, frozen, or pickled in vinegar. They are almost certain to be in any home vegetable garden.

9.

Beans

Of all the Leguminosae family of vegetables, the horse bean ranked first in ancient times. Today few care to eat these coarse broad beans because we have so many more tasty kinds. If Americans grow them at all, it is usually for fodder or a cover crop.

In many Mediterranean countries broad beans have been enjoyed since the beginning of time. They were one of the main foods in Assyria, Phoenicia, Palestine, and Egypt. Like lentils, they were ground into bread flour and stewed in pottage. Five thousand years ago, Egyptians placed them with funeral offerings.

In some countries the kidney shape of beans prejudiced people against them. Besides objecting to their resemblance to a part of the human body, scholars complained that no one could interpret the meaning of a dream he might have after eating bean pottage. At that time dreams were more important than diet, for through them the future was foretold.

Romans had great respect for the homely bean, and the celebrated family of Fabius was proud that it bore their name, derived from the Latin *faba,* in honor of ancestors who had cultivated fields of beans.

Two other famous Romans owed their names to vegetables—Lentulus to lentils and Cicero to peas.

Romans had a special ceremony at a certain time of the year when the head of every household took a solemn oath, pledging faithfully to perform his domestic duties. As he pronounced the words he flung black beans over his head nine times. Beans were also tossed on graves and eaten at funerals. They were thought to contain the souls of the dead and were sacrificed at a festival, accompanied by offerings of bacon. It may have been from this custom that the idea of eating pork and beans together originated.

Both Greeks and Romans used beans for voting at elections, white beans representing one kind of ballot, black beans another. Beans were used in a similar way at trials, the white one signifying that the prisoner was not guilty, the black that he was condemned. Thus in the beginning blackballing may really have been "blackbeaning."

Romans thought that beans mixed with goods for sale at ports would bring luck to the seller. When a farmer sowed a field he brought home a few beans with him, believing that this would ensure a good harvest. This was purely superstition, but farmers already knew that beans had a magical effect upon fields. They discovered that if they plowed bean vines into the earth just as they began to bloom, the soil was improved. Today we understand why. Leguminous plants, when grown in contact with certain bacteria, form small knots, or nodules, on their roots. These are traps for nitrogen, which is needed for plant food.

In the second century before Christ, Cato the Elder, a Roman statesman and agricultural writer, declared that growing field beans, lupin, and vetch, and plowing them under, was as good for the land as spreading it with manure. This knowledge was important to early civilizations. Not having easy means of transportation to other places, colonists expected to stay settled in one spot for generations. When their farms no longer produced good crops, they could not pack up and move on. They did not understand that the land lost its fertility because its nitrogen was exhausted, but they knew they must find ways of giving back to the soil some enriching material. In biblical times the Hebrews tried to solve this problem by setting aside every seventh year as a period of rest, when no tilling was done. They hoped this would cause the fields to renew themselves.

The Greeks and Romans let each farm lie fallow, or idle, every other year. In parts of the United States agriculturists still use this system. Beside a field green with young grain lies another covered with dry stubble. The following year the stubbly field will be plowed and the first one allowed to rest.

This scheme worked well for the Greeks and Romans, but farmers had to own much land in order to let part of it lie idle. A better plan was to rotate crops. One year they planted a grain that might exhaust the fertility, but the following spring they sowed the same land with peas and beans.

Our chemical fertilizers were unknown to ancient farmers, but some placed *nitrum* on seed grain to make it swell. This must have been either potash or soda mixed with oil. Ashes, rich in potash, also were spread on fields. Later, animal manure was added.

Until the discovery of America the commonly used beans were two called by Latin names, the *faba* and *phaseolus*. *Faba*, the common broad bean, is more nutritious than wheat and contains much nitrogen. Because it makes excellent coarse feed for horses, it is called the horse bean. Probably it first grew south of the Caspian Sea and in northern Africa and was carried to Europe so early that it was used by the prehistoric lake dwellers.

Ancients employed the *faba* in many ways. Bruised beans boiled with garlic were considered a good cough remedy. Bean flour was a Roman cosmetic; women believed it would remove wrinkles and smooth their skins.

Legend says that Alexander the Great discovered the *phaseolus*, the green haricot or kidney bean, commonly used in the United States and often called the French bean. He tasted it on his campaign in India and liked the vegetable so much that he sent seeds to be planted in his homeland, so that he might have garden beans on his return. However, scientists think this may be just another of the exaggerated tales related about Alexander.

Europe did not know much about the garden bean until the sixteenth century. Although various kinds grew in India, the tender ones we are so fond of seem to be descended from the beans of South America. England received them from the Netherlands in 1509, when gardening first began to interest the British. This new vegetable must have been a pleasant change from the old broad bean, which ever since the Roman conquest of Britain had been one of the main foods of the country. The British used to say, "Shake a Leicestershire yeoman by the collar and you shall hear the beans rattle in his belly."

When Columbus landed in the West Indies, he saw that the American Indians had beans of more kinds and colors than he had ever seen before. Both he and other explorers who came later told of the fields of these vines cultivated in the islands. The new types were carried to Europe, but as late as 1629 our commonest bean was still a rich man's dish on the other side of the Atlantic Ocean.

Both navy and lima beans have been found in tombs in the Andes Mountains, showing that the ancient South Americans buried them with their dead. Limas usually grow in warm climates, so they were not known as far north as common beans until, in 1824, a navy captain brought the seed from Lima, Peru, to his New York farm. He gave the vegetable its name. At that time limas already were being grown in the Carolinas.

Slave traders making voyages between Brazil and Africa were the first to carry the new beans to the Dark Continent.

People of the Orient for 25,000 years have used the soybean, called the most versatile plant known to science. It provided butter, fats, and cheese for the Chinese, to a great extent taking the place of meat and milk in their diet. The young sprouts also served as a leafy vegetable.

Although the United States grows millions of bushels of soybeans annually, their main use is industrial. More than 250 products are made from them, including paints, adhesives, lubricants, soap, varnishes, diabetic and infant foods, salad and cooking oils, soups, stock feed, and fertilizer. Bread can be made from soybean flour, and soybean sauce peps up dishes served in Chinese restaurants.

Soybeans and jack beans are commercial sources of enzymes—delicate substances, mostly proteins, which are used in chemical and biological research. Enzymes are also obtained from horseradish, sweet potatoes, mushrooms, and tomatoes.

Of the many kinds of beans in the world, the ones we know best are those that plant breeders have developed from the American *phaseolus*. Some have yellow pods, some green ones. When allowed to stay in the fields until old and dry, these pods give us hard white navy beans.

The American taste for Boston baked beans comes from an Indian custom of cooking this vegetable in earthen pots buried in hot ashes. The colonists copied this dish, as they also did the Indian succotash, a mixture of corn and lima beans that has kept its native name.

Children in Massachusetts colony ate so many beans that they changed the old verse which had been sung in England to:

> Bean porridge hot, bean porridge cold,
> Bean porridge in the pot, nine days old.

Beans were not just a nursery rhyme to these children; they often were the entire dinner menu. Wild turkey and cranberry sauce are so firmly associated with the first Thanksgiving dinner that we forget what foods actually kept the Pilgrims alive in those early years. Until the seeds they brought from England had time to grow, they ate pumpkins, beans, and corn day in and day out. To make the dull food more interesting, they sweetened the beans or cooked them with bits of fat meat.

Canning of beans, either fresh or baked, is a great industry in the United States. Because they are one of the world's most nutritious foods, beans have an important place on the menus of most nations. Canned or dried, they can be taken to the South Pole or to tropical deserts. The Mexican Indian thinks his small black beans, called frijoles, are so good that he will gladly eat them as dessert. The bean is truly a vegetable that feeds whole nations.

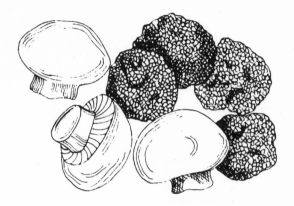

10.

Mushrooms and Truffles

Edible fungi are among the oldest vegetable foods in the world. The most common edible fungus is the mushroom. European countries also have the truffle, and in the region of the Red Sea there is a lichen that grows in small grayish lumps and which may have been the manna the Israelites found when they were lost in the desert. It is not the kind used as sweetening, but a dryer substance, probably carried out into the sands by windstorms.

Truffles were used as food before mushrooms, by the ancient Assyrians of Mesopotamia. These fungi are almost unknown in America except as garnishing for the tops of cans of imported goose-liver paste. The most curious thing about truffles is that they grow underground, clinging to the roots of young oaks, beeches, and other trees. Some are pink and about the size of a walnut;

others are black. No ordinary human could find these odd growths except by accident. In France, where truffles are raised commercially, dogs or hogs are used to locate them.

The ancient Greeks and Romans called these fungi "swine's bread" because pigs liked them. When the farmer wished some for his stew, he tied a cord to one of his porkers to prevent it from straying. Then he sat down to watch the animal, and when it rooted under a tree, the farmer knew where to dig for his truffles.

Early peoples could not explain the existence of edible fungi. Some believed that truffles were sown by thunderbolts; others said they were bubbles in the ground. Scholarly writers insisted that mushrooms were imperfections in the earth, and as proof pointed out that they rose from the soil without passing through any crevice or being held by roots.

Mushrooms pop up eerily in the warm spring and autumn rains, sometimes in circles that have long been called fairy rings. Since they seem to rise magically from the earth overnight, they used to be regarded as elfin. But there is nothing fairylike about their growth. Sometimes the spawn is imprisoned a year in the ground and often a mushroom remains in the button stage for weeks or months.

Mushrooms most frequently grow where horses have been tethered a considerable time. The spawn is not really seed but minute particles of the mushroom itself, which drop from the underside.

When one of these fairy umbrellas is pulled from the

earth, sometimes hairlike roots cling to the base of the stem.

Mushrooms were not cultivated until the Middle Ages, when Europeans discovered they could plant them by obtaining manured earth in which the tiny threads were embedded. If you want to buy mushroom seeds today, a nurseryman gives you a brick of rich earth or a cylinder of prepared organic material containing the hairlike threads, or mycelium, to be crumbled and spread in a dark place.

Mushrooms have to be grown in the shade. The mushroom caves of Paris are famous, and in the United States some of the best-known mushroom farms have been in abandoned mine tunnels. Horse manure, or an artificial compost, is spread on the beds and allowed to ferment before the spawn is mixed with it. When the plantings are not completely underground, they are in houses set down in holes in the earth, or in cellars. In Europe beds are sometimes planted out of doors and covered so that they will be dark.

Many varieties of mushrooms have been grown in Europe for several centuries, but only one type is raised in the United States. In central Europe, large sections of the public markets are stacked with mushrooms of every size and shape. Quantities are canned or dried for soup. They add flavor to meats and other foods and are good sources of vitamins and minerals.

But a wild mushroom picked by an unwary person may be a deadly poison. Anyone who is not an expert and knows beyond doubt which mushrooms are edible and which are not should never gather them, no matter how tempting they look.

Until the edible varieties were placed under cultivation, people who ate mushrooms often risked their lives. In spite of this fact, the fungi were so well like by Roman emperors that cooks who invented recipes for mushrooms or truffles could be sure of royal favor. Since the Romans were forbidden by law to use much meat, savory vegetable dishes were especially appreciated by them. This explains why rich men frequently sent to foreign lands for the vegetables served at their banquets.

Poisonous mushrooms looked so much like the edible kinds that a man never could tell the difference until they were in his stomach. The entire company at one Roman feast died from eating the deadly variety. The officers of Nero's guard also died from the same cause.

One day the Emperor Claudius dined on a dish of ragout that smelled deliciously of spices, mushrooms, and seasoning herbs. He was unaware that Agrippina, his wife, had entered the royal kitchen when the stew was on the coals and had slipped a few deadly toadstools into it. She was determined to kill her husband, whom she did not love, and had hit upon this scheme, which she thought never would be detected.

Claudius ate the ragout with his usual gusto, but soon after became seriously ill. However, something in Agrippina's plot went amiss. For a while the emperor was very sick from the poison, but in time he recovered. Suspicion fell upon Agrippina; her guilt was discovered and she was punished.

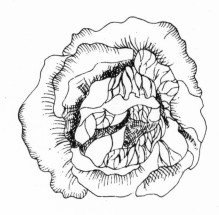

11.

The Cabbage Family

There was a time long ago in Rome when every illness had just one remedy—cabbage. If a doctor would not prescribe crushed cabbage leaves for wounds and dislocated joints, raw cabbage for gout, cabbage juice for deafness, and cooked cabbage for warts and weak eyes, he might as well fold his toga and move to another country.

For a number of years Romans were sure that in cabbage lay the secret of perfect health for the nation. This belief started when two doctors wrote books on its wonderful properties. They said that no disease would have a chance to lodge in the body of a person who ate a large quantity of cabbage, and that this vegetable would drive away any ailment that might already be there.

Since the time of Hippocrates, the father of medicine, plants have played a part in healing. Even before medicine was regarded as a science, home remedies were brewed from wild herbs, among them chicory and spinach. Usually doctors were expected to know how to make the most of the plants that grew near by. The Romans had plenty of cabbage, so this may have been the reason why they relied on its imagined properties.

Another of the feats they believed cabbage could perform was to prevent drunkenness. The Greeks, Egyptians, and Romans had this idea, for they thought the cabbage and grapevine were natural enemies, and that the former, being the stronger, would put up a good battle to keep a man sober. So if Father was dining out, he ate a plate of raw cabbage before taking leave of the family, in order to be sure not to return home tipsy. In that age, when wine was the principal beverage, a man might easily drink too much.

Wild cabbage must have been eaten by people more than four thousand years ago. Its native home is in the northern countries and different regions may have had different kinds, which were crossed as man carried the seeds in his migrations. Loose-leaved kale and collards are probably the oldest forms. Primitive tribes living around the Baltic Sea, lacking the comforts or the abundance of foods known in more temperate lands, placed great importance on their cabbages. The Scandinavians in particular welcomed them as a change from their fish diet.

The cabbage was brought from Europe to Egypt at an early date and was so adored by the people of the Nile that they erected altars to it and worshiped the vegetable as a god. They used its leaves for wreaths to wear at funeral banquets, entwining them with pea and bean vines and onion shoots.

One of the Roman names for cabbage was *caulis,* from *caul* meaning "head," from which cole and cole-wort, cauliflower, kale, and kohlrabi, the species mid-way between a turnip and a cabbage, got their names. There are about 150 members of the *Brassica* genus, which includes cabbages, cauliflowers, turnips, mustards, and related plants.

The ancient Greeks explained the creation of the firm, round heads with a story about two oracles who made prophecies that contradicted each other. Jupiter, god of the universe, was asked to decide between them, but was so puzzled that perspiration rolled from his brow in mighty drops. Where these struck the earth the first cabbages sprang up.

Storytellers were not as clever at inventing an explanation of the origin of kohlrabi, which they called the Corinthian turnip because it grew in the region of Corinth, Greece. The great difference between a turnip and a cabbage is that one is cultivated for its root and the other for its leaves. The kohlrabi behaved like neither. It seemed to grow upside down, with its turnip head just above the earth and its leaves waving on top.

The largest and most ambitious member of the *Brassica* genus is the tree or cow cabbage of France, which reaches from ten to fifteen feet in height. Such a coarse growth is not appetizing as food. The cabbages which interest gardeners are the delicate varieties, of which there are many. One can tell where some originated by their names, such as Brussels sprouts, Chinese cabbage, Savoy cabbage, and broccoli, which suggests Italy.

A twelfth-century Arab botanist was the first writer to tell of the "flowering Syrian cabbage" that grew in many gardens at the eastern end of the Mediterranean Sea. This was cauliflower. Its original home may have been the island of Rhodes. Genoese merchants carried it to Italy and from there it spread slowly through Europe. It was brought to America when it was still rare in France. The British took it from Cyprus to England, where at the beginning of the eighteenth century it was grown in small quantities as a luxury.

Broccoli, also a kind of flowering cabbage, was known in England in the sixteenth century and was referred to in the menus of Queen Elizabeth's time as "brawcle." Although English settlers brought a few seeds to Cape Cod in colonial days, broccoli was not generally accepted by Americans. Some Italian farmers in California grew a little for their own use, but it was practically unknown except among immigrants from the sunny Mediterranean country.

One day in 1923 an Italian waiter mentioned to his employer, who owned a fashionable New York restaurant, that his countrymen often went to some rubbish dumps at nearby Flushing and gathered broccoli they found growing wild. He said the sprouting flower buds made an excellent dish. The proprietor asked him to bring some in and proposed creating a special sauce for them and putting them on his menus. Patrons smacked their lips over the new food and from that time on its popularity spread. Now the vegetable is grown as an important crop in some districts. Frozen, it is available the year round. Most of us know only the green variety, but there is also another kind, with purple buds.

As the Italians had their favorite cabbage, so had other nations. The Spaniards prefer the red kind, and the French like curly Savoy.

Brussels sprouts were developed by farmers of the Low Countries. This vegetable, which was not known before the middle of the eighteenth century, is a cabbage-like plant with a high stalk along which the sprouts form.

China had pe-tsai, or Shantung cabbage; it is often confused with lettuce because it resembles romaine. Although it is naturally a green plant, Orientals prefer it blanched. Pe-tsai is the commonest article of food in their diet.

Cabbage is a plant for all seasons. In their first calendar the Saxons called February "sprout kale" because this was the month when the new shoots on the old stalks were ready for use.

Since the Romans considered the use of cabbage and turnips extremely important, they sent their soldiers on wars of conquest armed with seeds as well as swords. They were prepared to feed their legions when they settled in distant conquered lands. Therefore the Roman invasion of the British Isles was also an invasion of turnips and cabbages. Although the cultivation of turnips was neglected except in monastery gardens, cabbage and colewort took their place among the principal foods of the Britons, who believed, as the Romans did, that cabbage had curative properties, and many a Briton tried to cure a stiff neck with poultices of cabbage, goat's milk, salt, and honey.

Cabbage was well suited to the boiled dinners of the English. They had a native plant of the *Brassica* family somewhat resembling cabbage. This was sea kale, England's commonest vegetable, but unknown in most other countries. In ancient times the wild Britons and Scots searched gravel and sandbanks along their coasts for the tender shoots of this plant, which has remained part of their diet through the ages.

The world's most celebrated cabbage dish is sauerkraut, or pickled cabbage. The recipe is neither Roman, German, nor English. It was a dish the Slavic Russians learned to prepare after observing their Tartar neighbors. Wherever cabbage grew, sauerkraut sooner or later was made from it.

The botanical name of cabbage, which is *Brassica oleracea,* is very appropriate; *oler* is the Latin word for "smell."

Cabbage contains much sulfur, just as onions do, and for this reason some persons do not digest it readily. Cabbage that has been cooked too long loses its taste, color, and vitamins. When eaten raw it is rich in vitamins A, B, and C, also in iron and calcium, which are needed by the body for blood and bone building. All members of the cabbage family, including mustard greens, contain these valuable substances.

Since cabbage is a simple plant for seedsmen to experiment with, more changes have been wrought in this vegetable than in almost any other. The seeds were easy to carry from place to place, so they were among the first kind primitive man strewed in his clearings.

Most types of cabbages we use today were developed or improved in the Netherlands, Denmark, Germany, France, and England. When they were brought to the United States by early settlers they did not always thrive because of the climate, so American growers set aside the individual plants that had the most desirable qualities. The young plants were set out in the fall and allowed to winter in the field, for they are perennials and do not complete a life cycle in a single season. In the spring cuts were made in the sides of each cabbage head to enable seed stalks to push through. The ripe seeds from each stalk were saved and kept separate and planted the following year. This process was repeated year after year until the finest types were obtained.

Growers call this "breeding by selection." By means of it many kinds of cabbages were developed to satisfy various demands. In some parts of the United States farmers wanted those that matured early. In other places those maturing late were desired, or fewer outside leaves, or heads that were rounder, smaller, or more compact.

Some American growers found they could make more money by producing cabbage seeds to meet these special demands than by growing the vegetables for market. For this reason many Dutch and English on Long Island became the first breeders of American varieties of cabbage.

In California the growing of cabbage and turnips dates back to the gold rush of 1849, when some of the argonauts found there were other means of accumulating wealth than mining it. At that time there was a great lack of food containing necessary vitamins, and the miners were suffering from scurvy and other diseases of that nature. Vegetables then brought fabulous prices.

The Spanish Californians had been growing only onions, potatoes, red peppers, and beans, except for the produce of irrigated gardens inside the mission walls. Prosperous families brought their vegetables from miles away without knowing that they had some of the world's best garden soil outside their door.

Seeds were easily carried in the scanty baggage of gold seekers, and some people were surprised at the results obtained in their small plantings. Portuguese, Italian, and Chinese immigrants, mostly, got busy and started market gardens on a generous scale. Later, opening of railroads from the East encouraged them in growing important vegetable crops of many kinds besides the first cabbages and turnips. As the years passed, green eatables began to flow out of California by the carload and the trainload—and now by the planeload.

12.

Turnips

Long ago the king of the little country of Bithynia in Asia Minor was making war on the Scythians of southern Russia. It was winter and he was far from his beloved Sea of Marmora. The king did not care for the food of the country; he longed for a small fish like a herring, which is caught in his home water. So often did he speak of it that his cook tried to think of a substitute with which to tempt the royal appetite.

One day the cook had a happy idea. When the king and a guest sat down to the evening meal, a servant brought in a platter on which reposed a small white fish, fried in oil and salted and powdered with black poppy seeds. Was it possible? Was the king truly gazing upon one of the fish from the Sea of Marmora?

The monarch and his guest devoured the dish and praised its wonderful qualities. Everybody was content.

No one asked the royal cook where he had obtained the dainty. Only he could have answered that the "fish" was an imitation he had carved from a large turnip.

Today turnips do not grow so big, nor do cooks know how to give them the flavor of seafood. But they have been used to deceive people in other ways when edibles were scarce. Twice in the seventeenth century British grain crops failed and bakers had little flour with which to make bread. It was then they discovered that by kneading boiled turnips into an equal amount of wheat flour they could make excellent loaves.

The turnip was probably first known in Turkestan and Iran. Its ancient home may have covered all of the Soviet Union from the Caucasus Mountains to the Baltic Sea. It was used in prehistoric times in northern Europe and was among the first vegetables carried to China from the West.

Romans valued the turnip more highly than any other vegetable and said it was next to wheat in importance. When farms yielded more than the families could eat, the surplus was fed to the cattle.

The Swedish turnip is variously known as the rape turnip, the rutabaga, or the Swede. It is either yellow or white in color.

Although turnips are considered a plain food, nevertheless there was a Russian czar who was content to eat them raw all day long. Laplanders would exchange a whole cheese for a single turnip. The British put the likeness of the turnip on medieval coats of arms to represent a person of good disposition who gave to the poor.

They also praised it as a remedy for frozen feet or painful joints.

When the turnip grows wild it is useless as food, and if it had not been cultivated, we would never have known how good it can be. In places where the land is poor, these roots, with their generous store of starch, are especially appreciated, and the tops make tasty greens that provide vitamins and minerals. The Germanic and Gaulic peoples, who knew few vegetables besides turnips and cabbage, valued turnips as a precious food until the potato was introduced into their lands. Then they forgot the virtues of the turnip and that the two vegetables had many similarities.

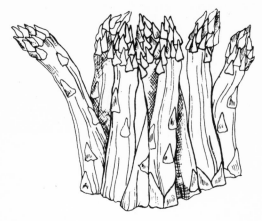

13.

Asparagus

In the second century before Christ, Roman farmers were watching with interest a new vegetable that had been brought from marshes and transplanted to kitchen gardens. It was asparagus. The name came from a Latin term meaning a young shoot before the leaves unfolded, for this was the part of the plant one ate.

In those days asparagus shoots were much longer and heavier than they are now. One alone must have been a meal, since Roman farmers grew them so big that a single stalk weighed three pounds. Travelers said that in Libya there were even larger plants with stalks twelve feet tall.

Asparagus is a kind of lily that first grew wild in damp regions and served as forage for cattle. It was common in the North Temperate Zone in Asia and Europe, but until the Greeks and Romans discovered that the tender shoots rising from the root stalk made a pleasant dinner dish, little attention was paid to it.

The sprouts can grow as much as eight to ten inches in a day and must be gathered at exactly the right time if they are to be eaten when their flavor is best. Pickers on modern asparagus farms begin work at daybreak and watch for those stalks that have as much green showing as possible but on which the heads have not begun to open. Sometimes workers pick over a field a second time during the same day.

Scarcity of labor and high costs are causing many growers to use machines for cutting. These take sprouts of all sizes. Now scientists are trying to breed asparagus plants that will all grow at the same speed and be equal in quality, to be more adaptable to mechanical harvesting.

At first only country folk in Mediterranean lands gathered the wild shoots for food, until it was rumored that the lacy plants that grew from these shoots were good medicine. People declared that they would cure toothache and eye trouble, and that when mixed with oil and used as a liniment they were protection against bee stings. Wise men said that asparagus was better for the stomach than any other herb. We know today that this food plant contains vitamins A, B, and C, and also calcium, phosphorous, potassium, and iron.

The delicate shoots promoted good appetite, and the Romans regarded them as the best dish with which to start one of their three-hour-long dinners. Emperor Augustus especially liked asparagus.

A Roman dinner at that time was quite amazing.

There might be five courses, which started with such appetizers as sorrel, lettuce, pickled cabbage, gherkins, radishes, and mushrooms. Next the servants brought in thrushes cooked with asparagus. Then there were meats, sauces, and pastries—until the guests leaned back on their couches, unable to eat another mouthful.

When Rome was in its glory the great authority on growing foods was a statesman named Cato the Elder. If he had lived in our time, he probably would have edited a farm magazine. As it was, he wrote a book in which he counseled the Romans on agricultural matters —how to press oil from olives, how to dress farm slaves economically, how to choose land for raising grapes.

Before he completed this volume Cato heard about asparagus, so he immediately decided to grow some. Until that time he had urged farmers to sow mostly turnips, cabbage, and beans, but his asparagus bed gave him so much pleasure he could not bear to end his book without praising the new vegetable. He told how to secure wild asparagus seed and prepare the ground for planting. After that every Roman who wrote a book on agriculture recommended asparagus, which took its place on rich men's tables as one of the luxury vegetables.

During the so-called Dark Ages, the early Middle Ages in Europe, knowlege of some of the more unusual food plants was almost lost and only the commonest and most easily cultivated were grown.

In the fifteenth and sixteenth centuries, Europe awakened again and people discovered they were missing many of the good things of life, among them asparagus. They secured seeds, perhaps from southern Russia, where the plant grew wild along rivers and near the seashore, and sowed them in moist valleys in northern Europe. When this vegetable reached England and was served in salads to Queen Elizabeth, lovers of good food spoke with respect of "sparrowgrass."

Across the Channel in France, common folk complained of a greedy king who wanted to keep all of the asparagus and artichokes for his court banquets. These dainties for the rich and noble were in time to find a place on every man's table.

Since 1896 asparagus has been grown and canned in great quantities in California. The first attempt to preserve this vegetable was made on Long Island in 1864, and gradually a technique for handling the shoots was developed. Those for market are set upright in crates. Those to be canned are brought in the same day from the fields.

14.

Artichokes

Roman aristocrats were great banqueters and enjoyed the best foods. The hard-working farm laborers, however, lived mostly on beans and turnips. Peasants thought so well of turnips that it seemed to them ridiculous when extravagant men ignored this native food and sent to Carthage in North Africa and Cordova in Spain for fancy thistles to serve at their feasts.

Pliny, the historian, believed that the commoners were quite right. "Those thistles," he wrote, "cost the Romans six million sesterces. Our countrymen are fools to serve vegetables which asses and other beasts refuse for fear of pricking their lips."

Why didn't the Romans, if they must have such peculiar food, eat thistles that grew in their own country?

It happened that those around Carthage and Cordova were wild artichokes quite different from the common thistle. The young stalks, when boiled, made a tender vegetable, but the choicest portion was the heart of the bunchy flower buds.

Much as the Romans delighted in eating this plant at their luxurious banquets, they did not trouble to grow it themselves, and not until the fifteenth century were artichokes cultivated in the domestic gardens of Italy.

When the Greeks were introduced to the fruit of the Carthaginian thistle, they named it *kaktos*. We borrowed the name in English but gave it to a different family of prickly plants, the cactus. The Romans, however, had another name for the artichoke. They said its spines were like the teeth of Cynara, the dog of mythology tales, so when botanists classified the vegetable, they labeled it *Cynara scolymus*. There were others who said that *cynara* comes from the Latin *cinis,* meaning ashes, and the plant was so called either because of the ashy color of its leaves or because ashes were used to fertilize it. None of this explains why today we call its fruit the artichoke, from the French word *artichaut*.

The Romans never attempted to eat the fleshy part of the leaves, only the base of the flower and the young stalks. They recommended seasoning them with vinegar and butter. They also preserved the hearts in vinegar and honey. Sometimes Greeks and Romans chewed the root, believing it would strengthen the stomach. Others warned that this practice would injure the voice.

Medical prescriptions called for artichokes as a remedy for jaundice or coughs. The juice, when pressed from the plant before it blossomed, was supposed to restore hair to bald heads.

The artichoke took kindly to garden cultivation, which changed it greatly. The farther north it was planted, the better it thrived, and the more the vegetable improved in taste and size. In 1548 it was introduced into England. Sooner or later British cooks of that period placed every food in a pie, so after a while they devised a pastry with baked artichoke hearts as filling.

Although we eat parts of the flower bud of the globe artichoke, there are still some persons who prefer the stem and leaf midrib of the plant, which are available in some markets, especially in California.

In certain parts of Argentina and Uruguay there were formerly many miles of prickly wilderness that had sprung up from the cardoon thistle (*Cynara cardunculus*), a close relative of the artichoke. Somehow, its seeds had reached South America and the plant had run wild. It became a pest that no Argentine rancher liked to see crowding his fields. Yet the cardoon is a poor relation of the cultivated plant upon which California growers lavish such tender care.

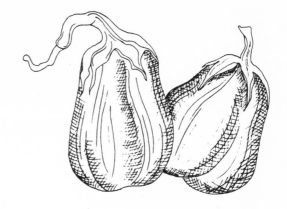

15.

Eggplant

When John Milton wrote *Paradise Lost,* he told of the fallen angels who wandered by the Dead Sea, hungry and despairing. They were in a land where smoke emerged from vents in the earth, bituminous pitch lay in pools on the ground, and the water of the sea itself smelled foul and was bitter salt.

When the angels saw luscious-looking purple fruits growing by the shore they gathered them greedily, hoping to refresh themselves, but as their teeth penetrated the smooth, shining skins, the pulp crumbled to ashes.

Milton based this tale on history and fable. It was told by Josephus, the ancient Jewish historian, who said he had seen the beautiful "apples of Sodom," which vanished in smoke when they touched the lips.

This mysterious Dead Sea fruit was purple eggplant,

and the legend of ashes within the shining skins was based on fact.

Long ago when people did not understand a phenomenon of nature, they explained it in a mystical manner. The plant with purple fruits grew near Sodom, one of the biblical cities upon which the Lord passed dreadful judgment because of its sinful inhabitants. When the town was destroyed, people were ready to believe almost anything about the fearsome region in which it lay. The smoking holes, the salt water, and the pitch certainly were part of the scheme of eternal punishment, they thought. It seemed fitting, too, that the fruits of the country should be only fair to look upon and not wholesome to eat.

Simple folk of biblical times did not know that the eggplant turned to ashes because a tiny insect caused it to decay and form a powdery substance inside while the fruit still appeared to be in good condition. Farmers of the region later learned more about how to deal with pests, and when the Arabs ruled Palestine, their ingenious gardeners placed the mysterious, detested pitch around the lower parts of plants to combat insects.

In the same land today, eggplant grows beautifully and is a favorite vegetable. We would scarcely recall its ancient history were it not for the botanical name of one variety, *Solanum sodomeum,* for the lost city of Sodom.

Throughout the Middle East, eggplant is extremely common. This vegetable was eaten in India four thousand years ago, and it may have originated there. When Chinese caravans began journeying over the mountains to Turkestan, early Oriental travelers reported to their emperor that the eggplant of western Asia amazed them. Instead of being small and round like that which the Chinese cultivated, it grew in enormous purple fingers, inside of which was firm yellow pulp full of seeds.

Although oyster plant (salsify) tastes like oysters, eggplant does not have the flavor of eggs. Its English name was suggested by the shape of the white variety, which hung on the bushes like "vegetable eggs." There are eggplants of other colors—yellow, ash, green, and brown. In the United States the purple species is preferred. In French this variety is called *aubergine,* probably from *brinjal,* one of its ancient names in the East.

Once the fruit was known as the mad apple, because the plant resembled the poisonous mandrake, then believed to cause madness. They are closely related; both belong to the same family as the tomato and the potato.

Europeans in the Middle Ages did not realize that eggplant could be eaten. The Italians probably learned of its usefulness through Turkish merchants and in the fifteenth century began serving it with their meals. The Turks had a dish called "fainting priest," made of eggplant stuffed with pine nuts, so delicious, they said, that it caused a priest to faint from pleasure when he sampled it.

Spanish colonists carried eggplant to the West Indies and from there it went to other countries in the New World. Meanwhile in England it was still unknown as food. It was grown in gardens for two hundred years as an ornamental curiosity before the British discovered that the fruits could be eaten.

Most of the Arab geographers and physicians who

wrote books during the golden age of the Moslems mentioned the eggplant. Arab doctors praised its qualities as early as the sixth century.

At that time Europe was going through the Dark Ages. While invading barbarians swept over the northern countries, the Arabs, who had prospered from their spice trade with the Indies, became accustomed to luxuries undreamed of by the Christian knights who fought them in the Crusades. These mailed warriors, fresh from the harsh life of British and Norman castles, where the winter diet was salted meat, boiled cabbage, and turnips, gazed with astonishment upon the food bazaars of Syria and Lebanon. Here they saw sugarcane for the first time, also artichokes, asparagus, truffles, dasheen, and cauliflower. Carrots, eggplant, endive, lettuce, and celery grew in the high-walled gardens that lay along every stream in the Lebanon Mountains. Aqueducts and monstrous waterwheels brought moisture to these fertile valleys, and Arab skill in cultivation gave quality to the humblest plants.

Scientists say that nothing contributed more toward increasing the agricultural flora of the Occident than the rise of the Arab nation. Before the time of Mohammed, the only real knowledge possessed by the wild Bedouins who founded the desert empire concerned the animals, plants, and stones of the peninsula on which they lived. In poems and folk songs they praised the qualities of the camels and horses they rode, and described the properties of the green things that grew around them.

As the Bedouins branched out and invaded new kingdoms, their leaders left the farming population undisturbed, for Moslem law held agriculture to be a noble work.

In the seventh century the Arabs, increasing in power, drove the Persian rulers away from Babylonia and its thriving fields of wheat, barley, and sugarcane, its vineyards, and date-palm groves. They absorbed tribes who had inherited the skill of the builders of the Hanging Gardens of Babylon and the farming and engineering talent of the Chaldeans and Assyrians, who turned the waters of the Tigris and Euphrates Rivers across the vast plains into a system of canals.

People of the Middle East, farming under difficulties since prehistoric times, knew how to make a small piece of fertile land support many persons. By feeding the coarser leaves of garden plants to camels and asses and drying the old roots for fuel and fodder, they wasted no morsel of vegetation even though it was unfit for human consumption.

Gradually, as the Moslem world spread from the Indus River to the shores of the Atlantic in Spain and North Africa, Arab agricultural methods and products followed the conquest. Plantations of sugarcane, cotton, lemons, carob and date palms marked their occupation of new territory. The Moors, invading the Iberian Peninsula, brought with them more than Oriental crafts and architecture; they planted for the first time in the West the luxury vegetables of the Syrian gardens—carrots, eggplant, lettuce, endive, spinach, radishes, cucumbers, and parsley. Furthermore, they taught the Spaniards how to select the types of crops best suited to the soil, how to irrigate, and what fertilizers to use.

These Arab conquerors of the Spanish coastal provinces were the most intelligent and successful farmers of that era.

The best book on agriculture written during the Middle Ages was the work of Ibn el Awam, a Moor who lived at Seville in the twelfth century. He was only one of many Arab writers who left a permanent record of his people's knowledge of plants, medicinal herbs, and farming methods.

Besides being practical farmers, the Arabs were lovers of beauty. They planted the homely turnip in artistic settings. Carrots pushed up among borders of flowers; fountains splashed musically beside onion beds; seedlings from Iran, India, or Africa thrived in rose-scented courtyards.

Of all the garden regions controlled by the Arabs, the narrow belt of coastal land along the Lebanon Mountains was among the loveliest. It had poured its riches into the treasuries of the ancient Romans, Greeks, and Phoenicians. Its steep slopes were indented with terraces, across which trickling streams were guided to irrigate orchards of oranges, figs, and olives, fields of sesame, and vineyards.

The principal occupation in this coastal region was supplying provisions for the caravans and ships which met at this crossing place of eastern and western trade routes. Much food was grown for voyagers who thronged the ports.

In this land of plenty the Arabs made no attempt to follow Mohammed's advice. The Prophet had counseled his followers to live simply, but Syria and the Leb-

anon had been a land of luxury ever since the Phoenicians built their trading empire. The nomad Arabs, who
were now its masters, developed a fine taste for strange
products, and, knowing that these delicacies were desired by other nations, they encouraged gardening for the
market.

When the Crusaders came down from the north, bent
on capturing the Holy Land, the Arabs fought as much
to save their gardens and aqueducts as to save their
religion. To them the knight was the infidel who would
destroy their farms and leave desolation where once had
been plenty. Those cavaliers who returned to Europe
from the wars related wonder tales of the sunny land
where fruits and food plants abounded and every high
wall hid some man's miniature garden of paradise.

As the ships of Genoa and Marseilles discharged their
loads of Christian knights and horses, their captains
stowed the empty holds with loads of merchandise to
carry back.

From the gardens of Ascalon, Tripoli, and Tyre went
bales of fresh produce, to lie side by side with perfumes,
spices, and silken stuffs of the East. The Genoese and
Venetians were leaders in this commerce, and it is easy
to understand why many of the luxury vegetables
common in Antioch or Aleppo were first grown on
Italian farms.

16.

Okra

During the Crusades an Arab writer, traveling under protection of the great warrior Saladin, visited Egypt and saw there a long hibiscus pod which dwellers in the Nile Valley called *bamiyah*. Occasionally the Egyptians ate the tender young fruits in salads, but the sticky mucilage inside the pod, which made the vegetable unappetizing to many, was wonderful for thickening soups. For this reason it became the most important ingredient of the Arab dish known as *gombo*.

In our southern states we have changed the name of this thick soup to gumbo, and the pods are known as okra. The plant resembles cotton and the two may be grown together as a mixed crop.

It is supposed that okra first was found in tropical Africa and was carried to India several hundred years before the time of Christ. It traveled west when slave traders made captives of African blacks and brought them to the New World.

The seeds of their favorite food must have come on the ships laden with frightened Negroes. On the return voyage the same vessels carried American provisions to the Guinea coast, and in this way corn (maize) reached the Dark Continent before it was common in Europe.

Practical-minded plantation owners in the Americas attempted to provide their slaves with food that was familiar to them, and the Negroes often were permitted to grow their own vegetables for private use.

As okra traveled across Africa with the Arab slave buyers and crossed the ocean to Brazil, Surinam, Martinique, and Barbados before the end of the seventeenth century, so it came to the United States with the earliest slaves.

Delicious gumbo soup and other southern dishes prepared with meat and okra probably have been adapted from recipes brought from the heart of Africa. Both Arabs and Negroes appreciated this food plant and spread its use wherever they lived, but Europeans as a rule avoid the slimy pods.

Creole ladies of the West Indies considered gumbo the choicest dish they could offer distinguished guests. This exotic food was the Arab couscous in an American disguise. Couscous, a concoction of rice or corn soaked in broth, to which potherbs, particularly okra, were added, was much liked by Gambia River Negroes in the seventeenth century. French settlers in Louisiana colony mentioned how delighted the slaves always were to have fresh vegetables for their couscous. The steaming plate of rice buried in chicken and okra sauce, which the southern hostess places on her table today, is a modern version of this recipe.

Persons who do not care for okra on first sampling the vegetable may acquire a taste for it. In the South the young pods often are boiled and eaten with butter or a condiment such as catsup, or they are served in salad. Okra seeds have been ground and used as a coffee substitute.

Okra pods taste best when they are young and soft, before the seeds are half grown; they are therefore plucked daily, almost as soon as they appear. Sometimes they are sliced in half and strung on cords for winter use. No Louisianan cook would think of being without an ample supply of this, her best soup vegetable.

17.

Fennel

In many parts of the United States a disagreeable, lacy-leaved weed encroaches time and again upon gardens. We pull it out and throw it away, for few Americans like the strong taste and odor of fennel.

But this plant, which is related to celery, was not disliked in ancient times, when the leaves were used in salads and sauces by European and Asiatic people. The Romans cultivated it extensively, and fennel was so commonly used that it was brought to the New World by early colonists and was one of the first things they planted in North and South America.

There are three kinds of fennel, one of them bitter and useful only in medical prescriptions. Another is sweet fennel, grown as an aromatic garden herb and for its seeds, the anise in old-fashioned cookies. The third is the vegetable often called finochio. It consists of broad, overlapping leaf stalks that grow in a firm, bulbous arrangement at the base of the stem. This species, *Azoricum,* was probably brought from the Azores islands.

Funchal in Madeira was named for the Portuguese word for fennel. When the island was discovered in 1420, the mariners immediately noticed an abundance of the plant on the spot where they landed and later built a town.

The fashion for eating fennel was started by the famous Medici family in sixteenth-century Italy, and the vegetable is greatly liked by Italians who have settled in the United States.

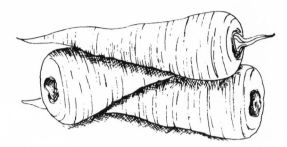

18.

Parsnips

Among the wild plants that grew in the valleys of western Asia were members of the useful parsley family, which includes parsnips, carrots, and celery. The parsnip is the least used as food today. Perhaps it was formerly valued because of its sweet taste. At any rate, when the vegetable was carried to Germany, peasants discovered that by boiling the roots they could extract a sugary juice from them. They made it into marmalade or used it as a base for wines and beer.

Wild parsnips were transplanted to gardens around the Mediterranean two thousand years ago and became popular with the Romans, who served them with honey and wine. The Emperor Tiberius was most particular about the parsnips he ate and sent far away to his provinces on the Rhine River to obtain the finest.

In some manner parsnips reached the British Isles; possibly they always grew there, for they were one of the few vegetable foods known to the people at the time the Romans invaded England.

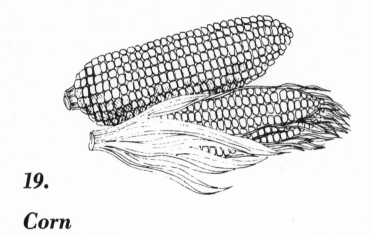

19.

Corn

The first Christmas dinner served in the Western hemisphere is said to have had for its main course roasted lizards and corn bread. Columbus and his crew dallied over the lizards and decided they could get along nicely without ever eating any more, but they were curious about the flat pads of bread that accompanied them.

No complicated dough was needed to prepare these odd pancakes, which were slapped on a hot clay plate and cooked over coals. A tasseled plant growing in nearby fields, called *mayz* by the Haitians, furnished the main ingredient of this bread; it was made of corn kernels which the women ground to paste with a stone roller.

Throughout the Western hemisphere these kernels were considered a celestial gift. Columbus saw fields of corn everywhere he went, and other explorers found the plant growing from the southern Andes to Nova Scotia. All American civilization, in fact, began in a cornfield.

Corn is so old that the Mexicans believed man was created from a grain of maize. Other Indian nations had similar legends. The Iroquois, for instance, said that in the earliest days of the earth the Spirit of Good brought birds and animals from the sun land. One of these was a crow that carried in its ear a kernel of maize which the Spirit planted in the breast of Mother Earth so that it might become the first grain of the Indians.

In Ecuador, people said there was once a great deluge from which only two brothers were saved. They climbed a mountain peak and waited until the flood had subsided. When they went down to seek food, two parrots helped them, bringing various edibles. After the birds returned several times with food offerings, the men captured one of them and it turned into a beautiful woman. She gave them corn seeds and showed them how to prepare a drink from it and how to cultivate fields of the grain. The corn maiden later married one of the brothers and from them was descended a powerful Ecuadoran tribe.

None of the myths gave any hint of how man actually obtained corn seed. Since no one had seen it growing wild, the simple people believed maize was a divine substance.

The sowing, cultivating, harvesting, and grinding of this grain controlled the very pattern of existence of the Mexicans. In their pyramid temples they worshiped a corn goddess, an ugly creature carved of rough stone, with ears of corn for hair. To her they offered the first of their harvest, and at certain seasons women danced before the plump idol, shaking their hair in imitation of the waving tassels in the fields. The Maize Mother meant to North America what Ceres, goddess of plenty, meant to the ancient Romans. Indians took their best ear to her temple, to be blessed and buried until harvest-time was over, when priests dug it up and distributed the grains as talismans.

Although the Mexicans held an eight-day fiesta for the Maize Mother in the season when the ears began to form on the plants, they were convinced that the seeds had been brought to them by Quetzalcoatl, the god of agriculture. They did not know that corn was older than their oldest god, and that the man who tamed it was an unknown hero, the greatest benefactor of his race.

Wild corn grew as a highland grass in Mexico fifty thousand years before men reached there. Tiny ears have been found in the caves the Stone Age people inhabited. At first they wandered in search of rabbits, rats, birds, turtles, small animals, wild vegetables, and, rarely, an antelope or the little wild horse that once roamed North America. More than 8,500 years ago the early Mexicans began gathering squashes, wild beans, avocados, and chili peppers. Next they added corn and several more vegetables. By 850 B.C. they were growing the corn in irrigated fields.

The Indians cleared land as best they could without tools. They had no plows to help turn over the earth, and their sole implement was a sharp stick, its point hardened in fire. The sower walked along, jabbing his staff in the earth at intervals and dropping in each hole one or two grains of corn from a pouch woven of rushes, which he carried slung over his shoulder. With his foot he pushed the soil over each planting before moving on to the next. He did the work neatly and economically, wasting no tiny plot of cleared ground, and his rows were as straight as if he had measured and spaced them.

Village life in Mexico began when the Indians had to stay in one place and tend their fields. While they waited for the corn to ripen, they planted gardens of other food plants. Their whole existence changed. They were better fed, and some of the tribes, such as the Aztecs, grew rich and powerful.

The Indians became more inventive, developing small shovels and a copper hoe, shaped something like a pick with a wooden handle. They devised a low, three-legged table of lava rock, with a roller for grinding. Housewives rose each morning at dawn and with this implement milled enough corn for the day's needs. The paste thus prepared was cooked in porridge or baked in flat pancakes, or tortillas. A steady diet of corn alone would have caused digestive disorders, so the Indians mixed other substances with it, particularly chili peppers. Tamales were one of their oldest recipes.

They learned to sweeten foods with the juice of green cornstalks. They designed ventilated cribs for storing the cobs, and used the shucks as fuel and as thatch for their roofs, or in making mats or tamale wrappings. They were so economical that they used every part of the plant. Even today no portion is wasted, for their barnyard animals eat the stalks that are left after the harvest.

This thriftiness also applied to the farmland. The Mexicans cultivated the steepest mountain slopes, wherever there was a patch of fertile soil, and when they were crowded on barren earth they built floating gardens, the only ones of their kind anywhere in the world.

The first of these was created about 1360, after the powerful Aztecs conquered their weaker neighbors in the high valley around Mexico City. Their cruel king, Tezozomoc, wished to impose an impossible task upon the vanquished Indians as an excuse to make them his slaves. He issued a proclamation declaring: "Unless you can furnish me with a field sown with maize, allspice, and pumpkins, which is light enough to float on water, I shall punish the tribes of this valley by making every man, woman, and child my slave."

Tezozomoc set a date by which the task must be completed, and thought his wicked deed almost accomplished.

The people, however, were not lacking in cleverness. They had lived so long beside the lake that they were skilled in fashioning watercraft.

On the appointed day Tezozomoc went down to the shore, expecting to punish his vassals. Amazed, he saw a field advancing toward him across the water, poled along by Indians. On it grew a garden composed of all the plants he had ordered. The ruler unwillingly admitted that the tribesmen had earned their freedom.

When Tezozomoc attempted to ruin the subject tribes by his unreasonable command, he actually did them a favor. The field they devised was built on a network of branches, shoots, and aquatic plants. On this raft was spread a bed of reeds covered with wet earth from the lake bottom, where the soil was much richer than on shore. Farmers saw that vegetation sprouted quickly in it and soon hundreds of rafts were floating on the lake, some complete with the owner's hut, a border of trees, and a patch of flowers. Usually each man-made island was forty-eight by fifteen feet in size and stood about a foot above the water. When the farmer tired of his location, or when he wished to visit relatives, he stepped into his dugout and paddled away, towing his field with him.

The Spanish conquerors, arriving in the sixteenth century, scarcely could believe their eyes when they saw gardens floating on the lake. The reports they sent home sounded like chapters from a fairy tale.

Years went by and the water in the lake became lower as it was drawn away in canals. Meanwhile the roots of the trees on the floating islands reached down into the soft mud on the bottom, gradually anchoring the rafts so that today they appear to be hundreds of tiny islands. Even though the owners can no longer pole their farms to the marketplace, these strange gardens still produce quantities of the vegetables used in Mexico City.

Throughout Mexico, life still centers around the growth of corn. No farm lacks the tall crib for storing the ears and no village is without grindstones for preparing tortilla dough. The Indian bread is made exactly as it was in the time of Columbus.

In his day there were two great civilizations in the Western hemisphere, one in Mexico, the other in Peru. Indians of the Andes Mountains had never heard of the Maize Mother and prayed instead to the sun-god for good harvests. On the night before the annual feast in his honor, Indian women prepared an enormous quantity of corn pudding and fashioned it into small round cakes the size of apples. This was the only time of the year that they used such bread, for ordinarily they ate the ears of corn toasted or boiled. Some Peruvian corn had kernels an inch broad and almost the size of chestnuts—big enough to nibble one at a time. There was also a pygmy maize growing on an island in Lake Titicaca, a kernel of which was considered a sufficiently powerful talisman to protect a man from hunger the rest of his life.

Peru has many types and colors of corn seldom seen in the United States, although some were known long ago to North American Indian tribes. The Navajos believed their first corn was sown from a blue ear that a hen turkey shook from her feathers and gave to them.

The Pilgrims saw the red, blue, yellow, and white corn the New England Indians stored for winter in grass sacks, placing them in holes covered with sand. Such a cache saved the passengers from the *Mayflower* from starvation when they ran short of food supplies. Winter was coming on and the colonists did not have enough provisions to carry them through the cold months, nor could their huntsmen bring in sufficient game to meet their needs. One day when they were most worried about food, one of the Pilgrims discovered an Indian storage pit filled with dried corn.

By the following spring the settlers were anxious to plant this new grain themselves, but did not know how to handle it. The Indians were better farmers than the white men, although their tools were poor and clumsy. Indian hoes had brittle clamshell blades or were fashioned from a moose's or a bear's shoulder bone, whereas the settlers had iron implements. But the Indians already knew a great principle of plant breeding—the selection of seeds. It was from the Indians that the Pilgrims learned to choose the best kernels for planting and to put a herring for fertilizer in each mound in which grains were sowed. In their first year of farming in the new land the Pilgrims had better crops than their best harvests on English farms. Corn grew luxuriantly with little preparation of the soil, and it had few enemies. The Indians taught the colonists how to keep the worms away from the plants and the importance of careful weeding.

The taste of corn was one to which the settlers were unaccustomed, but they soon learned to regard the green ear, which the Indians boiled or roasted on the coals, as a great delicacy.

Sweet varieties, which are most popular today, were then unknown because they had not yet been developed by plant breeders. The sugary taste was most noticeable in the tender young ears, which the Indians ate fresh. They saved the older ears for grinding into meal. When the whole community got together each harvest season to husk the corn, both whites and Indians considered it a good excuse for a celebration.

As the years passed, colored types of corn were grown less by the colonists. Sometimes a red ear turned up among the white ones and the finding of these furnished part of the fun at colonial husking bees. Many a pioneer romance began when the discovery of a colored ear gave its finder the right to kiss a pretty girl.

Although the Americans learned to appreciate this most important vegetable food of the Western hemisphere, it was a long time before Europeans liked it, and to this day many people do not care for the fresh ears that are one of America's favorite foods.

The Spanish conquerors were the first to take samples of maize to Europe, but the plant interested people so little that it traveled south and east instead of north.

The Moors had by that time been driven out of Spain, but many still carried on commerce with friends there. Through them the Moors early received seeds of corn and peppers and, recognizing their virtues, passed them on to neighboring Arab tribes. Before Europe knew much about their existence, these vegetables had traveled across North Africa and were growing in Turkish fields.

Exactly fifty years after the discovery of America a German botanist published a book containing a picture of maize. The caption beneath it said: "This plant has been brought to us only recently from Turkey and Greece, therefore it is called Turkish corn."

"Corn" was the general name for all grain in Europe. For half a century the plant was believed to be Turkish, until another botanist set the record straight.

"After growing it in my garden and trying it in bread," he wrote, "I am not sure of its virtues, but the barbarous Indians, knowing no better, think it a good food, but it is hard to digest and more convenient for swine than men."

Today's corn is tenderer than the common field corn of former years. It should be eaten soon after picking, when its flavor is best. The sweetness in corn was always present, but it has been increased through breeding. Each year, growers have selected the seeds of the sweetest varieties for planting and in this way have created better-tasting corn. The Indians knew that they could make syrup by boiling the stems of corn plants, and today we make use of their knowledge in the manufacture of corn sugar, a glucose often mixed with cane sugar in syrups.

Corn was not only a major contribution of the American continent to the food plants of the world, it also gave industry a valuable raw material. Its oil is used in making soap and it contains a gum that is a substitute for rubber. Dextrin from corn is used in library paste; the cellulose goes into insulating materials, and the residue into a plastic.

The corn products most familiar to us are breakfast foods, syrup, cake flour, hominy, starch, and cornmeal. Country people used to stuff mattresses with corn shucks and also use them for cigarette papers. The silk and ashes from the cob have medicinal value. Glycerine, liniments, dyes, alcohol, paints, varnish, oilcloth, and sizing for paper and cloth also have been made from this useful plant.

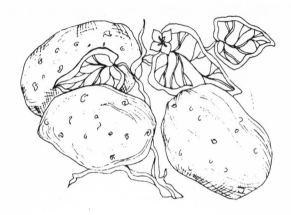

20.

Potatoes

The Western hemisphere's greatest gift to the vegetable kingdom was potatoes. It is difficult to imagine how people got along without them, yet they were among the last vegetables that inhabitants of the Old World learned about and appreciated.

At first the Spanish explorers of the Latin American countries paid no attention to the potato. Its story is something like that of the poor boy who made good. In the beginning this vegetable was a hard, scrawny little tuber, no bigger than a hickory nut, which clung to the straggly roots of a wild plant and struggled to keep alive in a harsh climate.

Between two and three thousand years ago some Indians in the Andes began to cultivate this tuber. These people lived in houses of rough stone, mud, and grass. They had no other products upon which they could depend for food, for nothing else would grow thousands of feet above sea level in the high ranges in Peru. The tight little brown tubers, with eyes like deep holes, were their only wealth. Some years the potatoes froze in the ground before they could be harvested, and if the Indians had not found a way to overcome this difficulty, they would have starved.

However, the mountain people invented *chuño,* or dried potatoes, which kept from season to season. If a drought or cold spell ruined the crop, the Indians ate *chuño* that year. When the harvest was good they had more variety, and could eat their potatoes boiled, roasted, or in soup. They needed no bread because the vegetable contains so much starch that a fine pastry flour can be made from it.

Preparing *chuño* required many days. The Indians spread the tiny potatoes on the ground for two weeks at a time when they were sure to be frosted each night. At the high altitude moisture evaporated from the tubers quickly, but to hasten this process the Indians tramped back and forth over the potatoes, bruising and squeezing them with their bare feet. When no more water could be trodden from the tubers, they were left to dry another fifteen to twenty days. They became light as cork and could be kept for many years. The Incas, who ruled Peru, used them for provision in case of famine and had vast quantities of *chuño* sent to stone warehouses where tribute was stored or army supplies were kept.

The Chileans also knew the potato, but its main home was in Peru, and as agriculture improved there, the people realized that some soils were better than others for growing it. If cultivated in the right kind of earth, it became a noble vegetable. After this discovery Inca Urko, famed as an architect and engineer, ordered hundreds of Indian porters to bring skin sacks of the best potato soil from Quito to Cuzco, where he built a great fortress. Its walls were fitted together without cement and yet are so tight that a knife blade cannot be pushed between the slabs. The fort was shaped like a half-moon and around it spread a field. On the eastern side Urko had his men unload the soil from Quito, and the mound they made of it was devoted to growing potatoes for the royal dinner table.

When the ruler wanted to know if a certain year would be lucky or unlucky for him, he called in a woman fortune-teller to read the future from his potatoes. She had a special way of prophesying by counting them in pairs.

Pottery makers of the Incas thought that vegetables were good models and many of their jugs and dishes were shaped like potatoes or squashes. Boys had whistles shaped like these vegetables, with little pipes rising from the top.

When the Spaniards reached Peru after the discovery of America, they could see no usefulness in the brown tubers. They were quite blind to the wealth with which Nature had blessed the Western hemisphere because they were seeking only gold and silver. To them the potato, one year's crop of which was worth more than all the treasure in the Incas' Temple of the Sun, appeared a very poor food. One Spaniard, setting down what he saw in Peru in 1538, described the tuber as "a kind of earth-nut which, after being boiled, is tender as cooked chestnut and has no more skin than a truffle and grows under the earth in the same way."

The Spaniards ate potatoes only when they had no other food. But there were some who discovered when they had no luck in finding gold that they could get rich—if more slowly—by means of the homely *chuño*. Miners had to be fed and dried potatoes were easy to carry. Why not take them to Potosí?

Potosí was a hill of precious metals in the midst of a barren highland in Bolivia. Here the Spaniards set up a mint to stamp out crude silver "cross money" and the pieces of eight celebrated in pirate chanteys. The treasure seekers who labored in this mining town had to eat, so certain shrewd men loaded llama caravans with bundles of *chuño* and sent them to the mines.

But the caravans did not carry potatoes outside of the high Andes. The vegetable was slow to spread beyond its homeland, and at first only mountain dwellers knew it. No one is sure when a Spanish galleon first loaded potatoes as part of its food supply before departing for the homeward journey to Europe.

Once the Spaniards tried using potatoes and *chuño* for ship's stores, they found them extremely valuable on long voyages. From then on they were a regular part of the food taken aboard vessels loaded with gold and silver on the Pacific shores of South America. Some adventurer must have taken a few tubers to plant in a Spanish garden, either as a curiosity or because he liked to eat potatoes. They went from Spain to Italy, and in 1588 the papal legate carried a parcel of them from

Rome as a gift to the Flemish botanist Charles l'Ecluse, who was director of the botanical garden in Vienna. The legate explained: "This is a new kind of truffle lately received from Spain. I know no name for it except that men returning from the New World say that the Indians call it the *papa*." L'Ecluse wrote about the plant, and tubers from the Vienna garden were carried to other parts of Europe.

Sir Francis Drake is credited with introducing potatoes into Ireland, but probably what he did was to give some to Sir Walter Raleigh as a present. Drake saw the Chileans cultivating this vegetable and probably added *papas* to his ship's stores. Or perhaps in his voyaging he captured a man who, having prospered in the *chuño* trade with Potosí, was carrying samples of his wares home to show the Spaniards.

At this time Raleigh was on his way to England after his trip of discovery along the shores of Virginia. Somewhere he must have crossed Drake's path and received a few tubers from him. When Raleigh gave the potatoes to his gardener with instructions to plant them on his estate near Cork, the story got around that Raleigh had brought the new vegetable from Virginia. This could not have been true, for the only tuber Virginia Indians knew was the openauk root.

The first potatoes planted in Virginia came out of a chest shipped from Bermuda, where the English had established settlements. When colonists were sent to found Jamestown, two large chests were filled with foodstuffs for them to plant and these were dispatched from the island. Some of the contents were from South America, including fruits, peppers, and potatoes, for it was mistakenly believed that the climate of Virginia must be the same as that in the Spanish colonies.

For a considerable time England regarded potatoes as a curiosity. At the end of the sixteenth century a barber-surgeon wrote a book in which he sketched a spray of potato flowers. He said the tubers could be used in pie, by adding to them two quinces, certain roots, a quart of wine, dates, egg yolks, the brains of three or four cock sparrows, sugar, rose water, and spices.

Long after Queen Elizabeth's reign potatoes were still so rare in England that they ranked with dates and oranges as dainties. In the seventeenth century they were cultivated in manor house gardens as exotic and curious plants.

Scottish ministers, hearing that some persons used the tubers as food, spoke against them in the pulpit. They declared that since potatoes were not mentioned in the Bible, they could not be good, hinting that they must be the forbidden fruit that Adam ate in the Garden of Eden. It may have been from this that the vegetable got its French name, *pomme de terre,* or "earth apple." The Germans called the potato *Kartoffel* because the papal legate who had presented them to Charles l'Ecluse spoke of them as truffles.

In spite of all that was said against potatoes, the Irish liked them, and in 1719, when a group left Ireland to found a new settlement at Londonderry, New Hampshire, they took potatoes with them. That is why we used to speak of "Irish potatoes" in the United States, because the Irish were our first potato farmers.

It was a long time before the tubers became popular because of the curious beliefs about them. Some persons said that if a man ate them every day he could not live seven years. Apprentices, on starting to learn a trade, insisted that they must not be obliged to eat potatoes as part of their food, because such a diet would shorten their lives. One colonial writer said that potatoes were not worth as much attention as horseradish.

Every spring American colonists who had grown the tubers burned those that were left over because they feared their cattle and horses might eat them and die.

It was not only in England that queer recipes for potatoes were concocted; the colonists cooked them with butter, sugar, grape juice, dates, lemons, mace, cinnamon, nutmeg, and pepper, and covered the whole dish with sugar frosting. One had to hunt to find the potatoes.

This vegetable had been carried from Vienna to Breslau and had been growing there for some time when Frederick the Great heard about it and ordered some potatoes to be planted in the Lustgarten in Berlin. Recently there had been seasons of poor crops in central Europe, and the Prussian king thought that if he could persuade the farmers to plant potatoes, they might be saved from hunger. His cooks invented recipes and he traveled from town to town giving royal dinners for leading citizens, introducing them to the new dishes, such as *Kartoffelpuffer,* a potato cake. Frederick also distributed bulletins on the planting and cooking of potatoes, but still could not induce the people to eat a vegetable they regarded with suspicion. Finally, after the famine of 1771, Frederick sent soldiers to compel farmers to plant potatoes.

Although the king believed they were good food and

forced his people to cultivate potatoes, he was not sure in his own mind that one could go contrary to the superstitious beliefs about the plant. He therefore decreed that potatoes must be planted only in the dark of the moon and must be dug up only at Michaelmas.

During the reign of Frederick the Great, in 1778, the War of the Bavarian Succession took place, in which Prussia and Austria opposed each other. Since there was little fighting and the main objective was to cut off the enemy's food supplies, the people nicknamed this war the Potato War.

Restaurants serve a dish called Parmentier soup, made from potatoes and named for the French horticulturist Antoine Auguste Parmentier.

When he was a young man, Parmentier was captured during the Seven Years' War and kept a prisoner at Hamburg. There he heard of the new vegetable that Frederick the Great was trying to promote and decided his own people ought to know about the potato.

During the year that Parmentier was released from prison, the French were suffering greatly from hunger because the grain crops had failed. The young scientist wrote a book explaining why his countrymen should plant potatoes. Everyone laughed. But Parmentier tried to convince them that the new food was good. He gave a dinner in Paris to which he invited Benjamin Franklin and other notables. The whole menu consisted of potatoes cooked in various ways.

Giving dinners to celebrities did not help him reach those who really needed potatoes. Food for the poor was a national problem, and in 1771 a French academy offered a prize for the discovery of some product to take the place of cereals in time of famine. This gave Parmentier a chance to exhibit the virtues of potatoes, and at last he caught the attention of King Louis XVI by winning the academy prize.

Parmentier sent a tub-of growing plants to the king, who signified his approval by wearing some of the flowers as a boutonniere. Marie Antoinette appeared at a ball with a wreath of potato blossoms in her hair. At once princes, dukes, and other noblemen rushed to obtain the blossoms so they might be in fashion.

It seemed that the only way to get people to eat the new vegetable was to trick them into doing it. So the king gave Parmentier about a hundred acres of land near Paris and a squad of soldiers to guard it. Parmentier planted his potatoes there and the soldiers patrolled the place as though they were protecting a treasure.

Farmers whispered among themselves that something valuable must be buried in the field. So at night people who had only recently made fun of potatoes sneaked into the field with sacks and stole some tubers. They took them home secretly, cooked them, and were surprised to find they tasted good. Soon many farmers were trying to grow them and Parmentier was chuckling over the success of his plan.

King Louis said to Parmentier, "Sometime France will thank you because you have found bread for the poor."

While the courageous Frenchman was waging his lone battle against hunger, the British were paying more attention to what was said of the desirable qualities of potatoes. A farmer wrote a letter to a magazine, saying that since the country did not have enough grain and meat, he would like to tell how he had kept his own family alive. He wrote,

> I plant every year four acres of potatoes. From them we make puddings and pies and boil or roast them for bread. In the morning my people eat milk and boiled potatoes. At dinner they generally have bacon or jerked beef and boiled potatoes. For supper we mash our potatoes and add milk and salt to make a pudding. Sometimes we have potatoes fried in fat that is left over from the bacon. And we also make pies with potatoes and meat. Many of my neighbors eat nothing but potatoes the year round and are very healthy and strong. For eight pence a man can buy a bushel of potatoes, which will keep him a fortnight if he eats nothing else.

From then on, more notice was taken of potatoes, and when Europe was experiencing a famine in 1771, the British common people were following the example of this farmer.

Potatoes are subject to various blights, the commonest of which is a parasitic fungus that causes leaves, stems, and tubers to decay. In a wet season the blight is worse and half of a crop can be lost. Such a disaster occurred in 1846 and changed the course of history. For almost a hundred years Ireland had depended upon this plant to provide the chief food for her peasants. When the crop failed that year, a famine swept the country and nearly a million poor people died of starvation. After two seasons of potato blight, the younger people decided that if they were to live they must leave the Emerald Isle, so many emigrated to America where, they were told, a person need not depend on one vegetable alone to keep hunger away.

This same blight spread to Germany and Poland, whose poor by then were also depending on potatoes for food. As a result, thousands of peasants emigrated to America. Thus the potato played a part in increasing the population of the United States.

Now that the world knows how to fight plant diseases, potato blight can be prevented by spraying and planting only blight-resistant varieties that have been developed.

The potato of today is quite different from the original tuber of Peru, which had such deep eyes that after they were cut out there was very little potato left. At first, botanists were not interested in improving potatoes because they believed they were poisonous, as they were closely related to henbane, jimsonweed, and tobacco. Farmers who did grow potatoes noticed that after a few seasons the plants became weaker from disease. It was not until various new kinds of fresh seed were imported from the Andes Mountains and a new stock was bred that we began to have sturdy plants suited to the United States. Since then, efforts have been made to develop a variety that will survive in the Arctic.

The potato is known as the jack-of-all-trades of the vegetable kingdom. In Europe, besides being used in pastry flour, the tubers are distilled into alcohol, which is used as fuel for automobiles and power machinery.

Of course the potato's prime use is for food. One of the most popular products is potato chips.

One night a chef in a famous resort hotel at Saratoga Springs, New York, was annoyed when a waiter came back with a dish of French fried potatoes prepared for a guest. "He wouldn't eat them," explained the waiter. "He wants them sliced thinner."

The chef lost his temper. Very well, if the gentleman wanted his potatoes thinner, he should have them as thin as mortal man could cut them. So the cook cut a fresh lot of spuds into the thinnest of slices and dropped them in hot fat. To his great astonishment, the potatoes that had been intended for a joke were an immediate success.

Rare is the farm in the United States that does not produce enough potatoes for household use. They are not only our greatest vegetable crop, but are more important than all others combined. The little dwarf of the Andes Mountains has become the giant of our dinner tables.

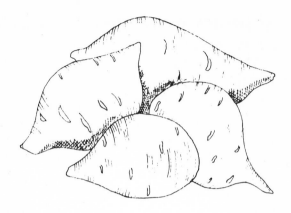

21.

Sweet Potato and Yam

The vegetable kingdom has one mystery over which scientists have long pondered. How did the sweet potato happen to be the only common food plant growing in both the Eastern and Western hemispheres when Columbus discovered America?

It is in the songs and legends of the Pacific Islanders that one probably finds the answer. They tell of Polynesian Vikings who braved the greatest of all oceans in outrigger canoes, moving from one South Sea archipelago to another until they left the islands behind and had bridged the gap to the mainland of Central or South America.

Hawaiian tradition relates that hundreds of years ago the vision of a new land was revealed to a learned priest who sailed a fleet of canoes across the Pacific and arrived in another country. He liked this place and journeyed there four times, but never returned from his last voyage.

The Peruvians have a story concerning the Inca Tupac Yupanqui, who learned from several strange merchants who came by sea to the port of Tumbes that their home was in faraway islands. The Inca thereupon made ready an expedition of balsa rafts to discover and conquer whatever land lay to the westward. More than a year later he returned, bringing prisoners, a treasure of gold and silver, and a throne made of copper and skins. He claimed to have taken possession of two islands.

The Peruvians were not navigators, but it seems certain that they knew of inhabited land lying to the west, for they instructed Spanish explorers how to reach Polynesia.

Where did the Peruvians obtain this knowledge unless the South Sea Islanders came to them? The distance to America from Polynesia is no greater than the route that the brown-skinned mariners commonly followed between Tahiti and Hawaii. Stories the ancient Mexicans and Peruvians told of tall strangers who landed on their shores are like the tales of the coming of the Maoris to New Zealand. And these Maoris are the people who carried sweet potatoes there from Tahiti.

At some period long ago, botanists believe, Polynesian voyagers reached the American shores. Recognizing the virtues of the sweet potato, they learned the methods of cooking, grinding, and storing this starchy food, and when they returned to their own lands, they carried the sweet potato with them to Polynesia. In this way, the plant eventually reached the continent of Asia.

Tahitians, like other Polynesian mariners, knew enough about the stars, trade winds, and currents to cross vast expanses of water without the aid of a compass. Aware that food was man's first concern no matter where he lived, they habitually carried bundles of plant cuttings. These were wrapped in coconut fibers, kept moist with fresh water, and sheltered from salt spray. As soon as the Polynesians settled on a foreign shore, they placed the precious shoots in the ground.

Like other races living in the hot regions of the Pacific and Indian Oceans, they drank almost no milk, ate little animal food except fish, and were mainly vegetarians. They were domesticators and cultivators of plants, their most important crops being taro, sweet potatoes, yams, sugarcane, bananas, breadfruit, and coconut.

The Polynesians performed an important service to agriculture by spreading nearly a hundred species and varieties of food plants among the Pacific islands. Many of the places were coral islets with almost no native vegetation suitable for food. In order to make these regions productive, the Polynesians hollowed out great pits into which they threw decaying coconut leaves and refuse until rich soil was created.

Adverse winds or curiosity may have driven the mariners far from the traveled course, and at some ancient date their canoes touched remote Easter Island, an outpost in the South Pacific rarely visited until recently. Here they set out the usual collection of cuttings, planting among other things the sweet potatoes whose descendants still thrive on the island.

One must imagine most of the chapters of this vege table conquest of the Pacific, for no written record exists. But sweet potatoes are mentioned in the oldest myths, chants, and rituals of Hawaii, Easter Island, and New Zealand. Planting and storing methods are spoken of and Hawaiian legends relate how the smallest sweet potatoes were peeled and dried for use on long voyages.

It is generally believed that Central or South America was the original home of the sweet potato, since a wild plant resembling it is found in eastern Guatemala.

Sweet potatoes belong to the morning-glory family. Although they have lovely pink or purple flowers, these never open fully and rarely produce seeds. New plants are grown from roots or cuttings of the stalk. In warm climates crops will spring up without having been planted. Cracks in the earth indicate when the potatoes are ripe and they can be harvested without disturbing the plant.

Sweet potatoes are a familiar item of diet in all tropical regions of the world. They have greater food value per acre than any other root and are considered potential sources of syrup and starch. In colonial days they were brought from the West Indies to serve as one of the main foods of Georgia and the Carolinas.

Filipinos, Chinese, and other Orientals eat sweet po-tato leaves as a green vegetable. The Japanese roast small pieces of the roots and sell them in the streets. Barbadians and South American Indians make an in-toxicating drink from them, and by combining lime juice with a red variety of sweet potato they have produced dyes ranging from pale pink to maroon. The Mexicans make a candy by boiling sweet potatoes in thick syrup.

The first explorers touching the New World realized that these roots were an important crop. Although they did not keep well, Columbus took a few to Spain to show to Queen Isabella. A few years later, when the first *papas,* or regular potatoes, were brought from Peru, the Spaniards began to call the sweet roots from the West Indies *papa dulce.* Thus the names of two quite different and unrelated plants were listed in English dictionaries as potato and sweet potato. Although both grow beneath the soil, the potato is an enlargement of the underground stem and is therefore a tuber, while the sweet potato is a fleshy root.

Columbus was not the only explorer who carried home samples of New World products. Sir Francis Drake or Sir John Hawkins must have taken the first sweet potatoes to Ireland.

There are two kinds in the United States, the dry northern type and the moist variety known in the South as yams. These should not be confused with true yams, which are members of another botanical family. At the end of the last century the Department of Agriculture imported true yams from the West Indies, but they have never been grown in commercial quantities in the United States.

Yams and sweet potatoes have fooled many wise men ever since the day when Portuguese sailors in Columbus' crew reported that the Caribbean roots were exactly the same as those which Portuguese ships had brought from the East Indies. Negro slaves in Portugal called them *inhame* or *igname,* meaning in an African dialect "something to eat." In English this was shortened to yam.

On his second voyage Columbus was accompanied by Dr. Diego Álvarez Chanca, an able botanist who regarded the roots with curiosity. "We are frequently visited by Indians," he wrote, "loaded with *ages,* a sort of turnip, very excellent food, which they cook and prepare in various ways. It is very nutritious and has proved of greatest benefit to us after the privations we endured when at sea." He added that they were mixed with another root and made into a kind of bread.

In Jamaica and some of the islands yams are so old that no one knows whether they were native or imported. One kind, the *yampie,* grows on a vine trained to grow on poles. It has a root like a sweet potato.

Eighteenth-century travelers reported that Barbadians seldom had bread, but instead roasted yams and ate them with butter or kneaded them into dough as they would flour.

We may well wonder if Dr. Chanca did not see both yams and sweet potatoes, although there are scientists who insist that no yams grew on the American continents until the French and Dutch brought them from the East Indies to their new colonies on the Caribbean shores. It is known that yams have grown in India since ancient times. Some day perhaps historians will be able to tell us exactly how and when these two nourishing roots were carried across the ocean.

22.

Dasheen, or Taro

There is another South American plant whose root resembles in appearance the sweet potato and the yam and which also may have made the same dangerous voyage across the Pacific in prehistoric times in a South Sea Islander's canoe. It is the dasheen, or taro, which used to be thought of mainly as a food of Hawaii, but which is now grown in the southern United States and is sometimes seen as an ornamental plant in water gardens.

Dasheen tastes like salsify. It is too acrid to eat raw, so the islanders boil or bake it, or make it into bread and puddings. It belongs to the arum or elephant-ear family and is the lovely variety of lotus that provided the ancient Egyptians with edible tubers, leaves, and seeds. Europeans called it the colocasia, and in the late Middle Ages it was known to nations around the Mediterranean Sea as an important food. Only when the potato took its place did the plant cease to be so highly esteemed. The Chinese, however, are one race that would rather eat dasheen than ordinary potatoes.

Taros were first cultivated in the United States in the seventeenth century to provide an inexpensive food for slaves in South Carolina. They grow only where there is a long, warm, frostless season.

23.

Jerusalem Artichoke

Although the North American Indians had neither the wizened little tubers of the Andean highlands nor the tropical sweet potatoes and taros, they had potatoes peculiarly their own. These came from a sturdy and widespread plant that in many parts of the United States is regarded as a weed.

Originally a native of the Great Plains and possibly coming from farther south, the plant spread readily and had reached the Atlantic Coast and the St. Lawrence Valley when the European explorers discovered it. They noticed that the huge yellow flowers of this plant turned their "faces" as the hours passed, so that all day long they looked toward the sun, like sunflowers.

It was the French explorer Champlain who, stopping at a small harbor on Cape Cod in 1605, was the first person to report the existence of this plant. He noticed that pieces of the roots, which the Indians cut and boiled with their soups and stews, tasted like artichokes. But it must have been an Italian or perhaps a Spaniard who named it *girasole,* or *girasol* (sunflower), and it was not long until its name, "*girasole* artichoke," had been turned into Jerusalem artichoke.

The vegetable probably was introduced into France by a companion of Champlain, and a few years later it reached England from the American colonies. It was an Italian named Colonna who first described it, in 1616.

Botanists have claimed that the native home of the Jerusalem artichoke was both Canada and Brazil, but in reality it belongs to the United States, and the plants that the early explorers saw in the St. Lawrence Valley and other places had been brought there from the Ohio and Mississippi basins. The Jerusalem artichoke grew spontaneously and needed little care. Indians ate it either raw or roasted, or in stews or soups. One species was cultivated by Indians of the Atlantic Coast to provide seeds for soups. Farther inland on the Great Plains grew another kind with edible roots.

Europeans were immediately enthusiastic over this Indian food, which is now better known abroad than in America. One reason may be that the plant is so aggressive that pioneer farmers in the United States treated it as a nuisance. Children of early settlers, however, liked to pull the roots and crunch them raw.

The United States never has planted any great quantity of this product, which is used mainly as forage for farm animals. But Europeans saw great virtues in it, and it often filled a need in times of war or famine. Since the plant always took care of itself and was not dependent upon cultivation, it was placed in out-of-the-way corners and harvested when wanted. Toward the end of the Thirty Years' War the people of Thuringia were thankful to have these wild sunflowers on which both men and animals were able to exist.

Of all the primitive North American root foods, the Jerusalem artichoke is the only one that was adopted by the white man. Pioneers had no desire to eat skunk cabbage, jack-in-the-pulpit, camas bulbs, water-lily seeds, clover roots, bitterroot, and other odd plants they saw Indian women gathering. Black moss, cooked in pits with hot stones for several days, was a typical delicacy of Indian cuisine. When ready to eat, it looked like jellied ink.

The Missouri River Indians cultivated corn and squashes and dried wild fruits and berries, but their favorite food was a kind of turnip they ate raw. These Indians often routed their buffalo hunts through places where the root abounded.

The Sioux gathered Indian potato and falcata beans, of which there were two kinds. The larger and more delicious ones were collected by meadow mice, and the Indians searched for these hoards, which often held a pint or more. Whenever they robbed the mice, they left handfuls of corn to take the place of the tasty beans they had stolen.

24.

Pumpkin and Squash

Once upon a time, North American Indian mothers used to tell their children, there lived a Corn Lady who had two lovers, the Bean and the Pumpkin. When she agreed to marry the Bean, he stayed in the maize fields and held his beloved in his embrace. The Pumpkin ran off over the ground to hide his sorrow. That is why bean vines climbed the cornstalks in the Indian gardens and the pumpkin vines wandered between the rows.

Pumpkins and squashes are the clowns of the vegetable world. The impish manner in which they swell large enough to cut into masks for Indian dances gave early Scottish settlers in the United States a unique idea. It was the custom in their old country for children to carry jack-o'-lanterns on All Hallows' Eve, the last night in October. The practice may have had to do with the lighting of fires at the harvest festivals of the ancient Druids or it may have originated, as the Irish declared, with a stingy man named Jack, who was barred from entering heaven because he was so mean. The Devil refused to let him into the place of eternal fires because Jack had tricked him so many times. Having no other choice, the stingy man was condemned to walk the earth with his lantern until Judgment Day.

The Scottish children believed in Halloween goblins and witches, so they made jack-o'-lanterns out of large turnips, put candles inside, and went forth to frighten the neighbors.

In the United States giant turnips were not so plentiful as they were in Scotland, but the Indians grew magnificent pumpkins, in which children could carve comical faces. Another nice thing about pumpkins was that they played an important part in the harvest season in the New World. Therefore, Halloween was both a night for spooks and a harvest festival.

So, early in the history of the nation, pumpkins became associated with one of our merriest holidays, and their worth as food was even greater in colonial days than now. The first settlers often had the pumpkin to thank for saving their lives. Someone grew so tired of eating this vegetable that he wrote:

We have pumpkins at morning and pumpkins at noon,
If it were not for pumpkins we should be undone.

Most of our pumpkins and squashes today are descended from those grown by the Indians. Four thousand years ago the Basket Maker tribes of Colorado and Arizona were using them for food. The cliff dwellers of New Mexico grew them, and the seeds, carried in trade to other tribes, were spread over a large part of the continent. They have been cultivated almost as long as corn. The Indians often related that pumpkins were brought by the Great Spirit, who came to earth in the form of a woman. She was supposed to have fallen asleep in a pleasant glade and, upon awaking, walked through the land, causing useful plants to spring up around her. Where she had lain tobacco grew, and maize rose up from her footprints. To the right and left beans and pumpkins flourished, and thus the needs of the Indian were supplied.

So many kinds of gourds developed that the Indians found other uses for them. They made mixing bowls and ladles, roof drains, pottery smoothers, ceremonial masks, and rattles from certain types they did not eat. West Indian Negroes stretched cords across bottle-shaped ones to make mandolins, which they called *bauzas*. Boys and girls learned to swim by using dried gourds for floats. People drank from gourd cups and sprinkled their gardens with water from gourd dippers. One variety furnished a dye, and the seeds of another were made into a cosmetic skin cream for Creole ladies in colonial days.

The Indians mixed the pulp of edible gourds with cornmeal and made it into bread. For winter use they cut squashes, dried and strung the pieces, and hung them up.

European colonists were surprised when they saw Indians put whole squashes in the fire, with no further labor to prepare them for eating. Here was food the English and Dutch had not known at home, and they wrote their relatives about the "vine apples" of the Indians. After a while they simplified *squotersquash,* the long Indian name, to squash.

"Pompion" was an old English word used to describe pilgrim or bottleneck gourds and other summer squashes that had come from Hindustan. The seeds reached England in ships of the East India Company after journeying the long way around the Cape of Good Hope. Among them was a variety shaped like a cucumber, which in the United States we call by its Italian name, zucchini.

Vegetable marrow is the common term in England for all kinds of pumpkin and squash, whether descended from Asiatic or American gourd families. Some variety grows in every temperate and tropical country and they have endless shapes, ranging from turbans to nutmegs, fruity-looking chayotes and warty crooknecks. In size they vary from the little scalloped pattypans of North America to the giant Asiatic squash known to reach 240 pounds in weight and honored by Chinese as the emperor of all vegetables. East Indians have a mythical tale about one which was so large that a man named Iaia used it as a casket for his dead son. He took the squash to the foot of a mountain, and one day, desiring to gaze upon his boy's face, he revisited the spot and opened the great gourd, whereupon whales and other monstrous fish jumped out of it. Iaia was so frightened that he rushed home to tell the neighbors. Four brothers who heard the story plotted to steal the gourd and fish in it for food, but Iaia followed them and when they saw him they dropped their burden and ran. The squash cracked in several places and water poured forth in rivers that inundated the earth and formed the oceans.

So much for legends. The closest connection the Asiatic gourds had with water was that they served as bottles in the ancient world. They became models for the wine bottles of today.

It is fascinating to watch gourds growing. The energy stored in their seeds is astonishing. The vine itself is so fragile that it may be destroyed by an early frost or a drought or a high wind, but if it does not meet with misadventure it spreads with great vigor, climbing fences, sprawling over sheds and taking possession of the garden. The baby squashes grow in strange shapes, and a farmer never can tell in advance if they will have crooked necks or mumps in one cheek.

The first colonists in America had to learn new kinds of agriculture from the Indians. They were unprepared to hew homes from the wilderness and they expected to encounter a different kind of weather. The British adventurers who first saw Virginia visited the coast in midsummer and gained the impression that the climate was tropical. Consequently the settlers founding Jamestown expected to grow such unsuitable products as figs, olives, and oranges. They brought instructors to teach wine making but had no one to tell them what to do when winter blanketed the hills with snow.

Captain John Smith had captured two young Algonquin Indians, and before long these youths were giving the English their first lessons in how to grow foods centered around a cornfield. Every Algonquin family cleared a field from the wilderness, usually by burning off the vegetation. The earth was turned over with a mattock or hoe, made from the shoulder blade of an animal. All the planting was done by the women. They sat on the ground and made holes in the soft earth with a plain tool about a foot long and five inches broad. In each hole they placed four grains of corn and two beans. The plantings were neat rows of mounds carefully hilled up, which the children helped to weed. As harvest approached, the Indians built tiny houses on platforms in the fields and someone stayed in them to frighten away the birds.

The Indians had in their gardens four varieties of corn, beans of several colors, sunflowers, various kinds of squashes, and the Mexican tea plant, with which they flavored dishes.

Although the Indians depended mostly upon field crops for food, they gathered everything tasty that grew wild. They dried fish and venison, fruits and berries for winter. Acorns, chestnuts, wild rice, and groundnuts were put away in baskets. Sometimes the women went to the marshes and dug a fungus called tuckahoe. They lived well, and before winter was over the white settlers were buying fish, game, maize bread, and pumpkins from the Indians. Many strange bargains were made—a copper kettle for eighty bushels of corn, two pounds of blue glass beads for three hundred bushels. Soon the colonists' wives were copying native recipes, such as cooking beans, corn, and pumpkin together. There was never enough food in any of the thirteen colonies for the newcomers to have variety, for they did not know how to use wild products. That was why pumpkin for breakfast was no laughing matter. "Let no man jest of them," declared one colonist, "for with this fruit the Lord was pleased to feed his people."

The time came at last when the settlers' gardens were filled with turnips, parsnips, carrots, cucumbers, watercress, leeks, onions, and seasoning herbs grown from seeds brought from England. Not until then could they say, "No, thanks, I've had enough squash."

When the colonists wished to make a real treat of pumpkins, they seasoned them with spices and put them in pie. Next to apples, many cooks declare, no other substance makes such satisfactory pastry filling as Hubbard squash.

Until 1798 the Hubbard squash had not been seen in this country. A ship's captain must have brought its seeds from an unknown place in the West Indies or South America, and some were planted by a Massachusetts farmer. He went into the town of Marblehead with produce and sold a squash to a Captain Martin. The captain liked its taste and saved the seeds for his garden, sharing them with a neighbor, Mrs. Elizabeth Hubbard.

When the squashes ripened, Mrs. Hubbard visited a seed grower, J. J. H. Gregory, and said, "Come to my house. I have something to show you in my garden."

There among the leaves lay dark-green squashes that turned up at one end. Gregory sampled the pulp and asked if he might cross the new squash with varieties he was growing. When Mrs. Hubbard consented, the seedsman showed his gratitude by naming the new vegetable for her. Since then plant breeders have changed the size, shape, and color of this squash, but it has actually been the forebear to a number of new varieties.

Today over 120 kinds of squashes grow in the state of New York alone. They have been developed from seeds native to both the Old and New Worlds, but by far the most appetizing are from the temperate part of North America. Some are direct descendants of the oldest squashes.

One day an archaeologist exploring cliff dwellings in Arizona entered a room that had never been opened before. It contained ancient pottery and household utensils, and in one corner were a few corncobs and a dozen squash seeds.

Out of curiosity, the man planted the squash seeds and watched to see if anything would happen. Eleven did not sprout, but the twelfth still contained a germ of life after many hundreds of years. From it sprang a healthy vine on which grew a warty green squash weighing nearly twenty-five pounds. The man saved its seeds to plant again. He had conjured out of the earth a genuine ghost of the vegetable kingdom.

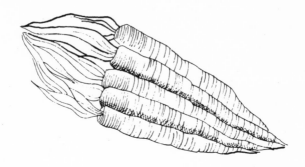

25.

Salsify

Salsify, an edible root of the sunflower family, resembles a small parsnip, and its distinctive taste has caused it to be nicknamed "oyster plant."

Italians call it *sassefrica*. The ancient peoples of Greece, Dalmatia, Italy, and Algeria knew it well. Possibly the Romans carried it to England, where the plant escaped from cultivation and grew wild.

Some salsify was planted in the West Indies in the sixteenth century, but the first time it was heard of in the United States was when an Italian doctor settled on a farm adjoining the home of Thomas Jefferson at Monticello. This physician and wine merchant, Philip Mazzei, had come to Virginia to be in charge of an olive- and grape-growing enterprise. When he arrived, he found the land unsuitable for the project. Thomas Jefferson came to the rescue by offering two thousand acres of his estate on the James River for the agricultural experiment. Mazzei remained at Collé farm five or six years, and during this time Jefferson found the doctor a first-rate agricultural instructor. After Mazzei's departure on a diplomatic errand, Jefferson hastened to employ the Italian gardeners the doctor had imported to care for the plantation.

To most Americans the name of Thomas Jefferson immediately suggests the Declaration of Independence. Few Americans realize that the great colonial statesman was above everything else an agriculturist. "When I first entered on the stage of public life," he declared, "I came to a resolution never to wear any other character than that of farmer."

He inherited a large estate from his father and added to it until he owned five thousand acres. Although only a small portion of his income came from farming, it was the calling he loved most to the end of his life, and nothing in his public career gave him so much satisfaction. He enjoyed vegetables, believing them necessary for man's health and comfort, and when he laid out the grounds at Monticello, elaborate steps were taken to provide for them. Workmen blasted rock and built walls to support terraces where native and imported seeds and cuttings were planted. Jefferson raised a greater variety of produce than is found on the best farms today, and his table was laden with foods considered luxuries in the colonies. His fame as a host spread so far that in time Jefferson's guests literally ate him out of pocket money and into a mortgage. His housekeeper on occasion had to provide fifty extra beds for guests, and slaves left their productive field work to serve food and drink at his bountiful board.

Far from being content merely to plant staple articles, he experimented with all sorts of cresses, lettuce, celery, cucumbers, asparagus, salmon radishes, Spanish onions, peppergrass, and other exotics. Until his meeting with Dr. Mazzei, Jefferson never had known Italians, and he watched eagerly the farming activities of his new neighbor. Mazzei shared seeds and shoots with him, and in 1774 Jefferson entered in his garden books his first planting of "salsifia."

Little of the vegetable was grown commercially in America until 1806, and most found in the markets today comes from California.

When horticulturists wish to know what vegetation grew in Virginia in the colonial period, they turn to Jefferson's records to read of the native tuckahoe, Jerusalem artichoke, wild pea, maize, pumpkin, and cymling squashes, which the settlers, copying the Indians, grew successfully. We know from these notes that Jefferson planted tomatoes before most Americans appreciated their qualities, that he had okra, four kinds of beans, and a host of other edible plants.

Jefferson knew where every tree, shrub, or vegetable patch at Monticello was located and where a bush had died or one was missing. He made note of the dates of the planting and ripening of produce, and each morning he rode over the plantation, inspecting crops and giving instructions. When he was in Washington, he sent home many kinds of trees and shrubs. The cart which his servants loaded with vegetables and fruits at Monticello for use on his table in the capital seldom returned empty. Jefferson sent back in it nursery stock to be set out. Even during his years abroad he observed farming methods, dispatching new seeds to America and making suggestions about their handling. He had a keen interest in inventions for the improvement of field work and was a forerunner of modern agricultural engineers.

He believed that the greatest service a man could render any country was to bring it new and useful plants. He had a special appreciation of the place of agriculture in national life. "Cultivators of the earth are the most valuable citizens," he maintained. "Agriculture is the first and most precious of all arts."

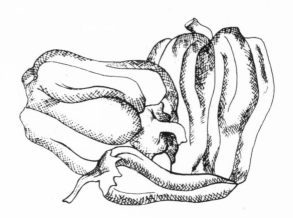

26.

Peppers

Before the coming of the Spaniards to the Americas, when the Incas were kings of Peru, one vegetable, the pepper, was so much a part of daily life that to be denied it was a severe hardship. For this reason, it was one of the things a Peruvian boy was required to give up during the ordeal of becoming a warrior.

Every youth of noble birth was obliged to prove his manhood and earn the right to wear the knightly sandals that were a mark of chivalry. He performed certain feats of strength, such as wrestling, running long distances, and staging mock battles in which he might really be wounded.

For the entire month while these tests took place the boy had to sleep on the bare ground, dress in poor clothing, and go unshod. He was allowed only the scantiest food and sometimes he fasted for several days at a time.

When he did receive nourishment, he was permitted no peppers, either green or red. They were denied him because no matter if his meal was only a dish of tasteless herbs, this piquant spice made it savory. The Peruvians put peppers in nearly everything they ate, so to deprive a boy of his *uchu,* as they called it, was like taking away his salt and sugar.

When peppers were not eaten raw or boiled, they were hung up to dry for use as seasoning. Tropical forests of the lowlands had other spice plants, but the Peruvians were content to employ only this one hot seasoning.

The peppers were of many varieties and shapes, including the small, hot chilies from which Tabasco sauce and cayenne pepper are made. An Indian child thought nothing of eating these straight, although strong men in other lands would hesitate to gulp down this most fiery of all edible substances. Primitive races do not have the delicate tastes of modern man, and the hotter the climate, the hotter the food they desire.

In the United States the hot chili pepper is used in small quantities. Our preference is the sweet bonnet or bell pepper, served in salads or cooked with meats. Spain and Hungary cultivate large fields of this pepper and dry it to make powdered paprika. As the bell pepper ripens it turns red, and is very popular as the sweet pimento we buy in jars and cans.

All of the green and red peppers belong to the *Capsicum* genus and are not related to the vines and bushes of Asia and Africa that furnish black pepper. Latin Americans know many fine differences of flavor between the various peppers and use all of them.

In honor of this greatly respected vegetable, the name Uchu was given to one of the founders of the Inca race. Uchu, according to legend, was one of eight brothers and sisters who came miraculously from an opening in the rocks near Cuzco. They were godlike, and one was chosen as the first king of Peru and another as the first queen.

These royal brothers and sisters, the legends said, taught the Indians how to plow, sow corn, and irrigate the land, and pointed out the plants fit to use as food.

As the Inca Empire grew strong and conquered more tribes, colonists were sent to each of the new territories to plant fields of corn and peppers and show the primitive subjects how to farm better. Since the Peruvians had neither writing nor money, they kept records of their crops by means of knotted strings. Maize, *chuño,* and peppers served them in place of currency. They had few tools and no domestic animals suited to pulling farm implements. Their plow was a pole with two wooden crosspieces. The plowman set it in the ground at an angle, then jumped on the crosspieces so that the pole was driven into the earth.

Seven or eight plowmen worked together in a field, all singing as they set their poles and jumped in unison. Women worked along the deep furrows, breaking up the clods, shaking earth from plant roots, and tossing weeds aside.

Certain rules for cultivation had to be followed because they all worked together. All of the land good for sowing corn was divided. One third was for the Sun, which the Peruvians worshiped; one third was for the king; and one third for the people. The religious domain was the first to be plowed, then the lands of widows and orphans, the aged and the sick, and after that the regular village lands. In order that there should always be enough food, each community planted more than it needed, so that in time of famine the granaries would be full.

The last fields to be tilled were those of the king, and there was great rejoicing during the time when the Indians were engaged in this work. They went to the fields wearing their best garments, and when they were through they staged a feast.

Skilled men planned how to use the land, for the mountains were steep and exceedingly high, whereas the valleys were narrow. There were sharp differences of climate as one climbed out of the hot tropics where cassava grew and entered the cold regions where only small potatoes struggled to keep alive. A Peruvian mountaineer, spreading his *chuño* to dry between snowstorms in the highlands, could look down upon steaming jungles far below him. In Peru the climates of places within a few miles of each other may be as different as those of Jamaica and Alaska.

The Andes Mountains have not worn down in soft ridges; they are jagged and rocky, with little soil for cultivation. The Incas had to bring garden soil to places where it was needed. They were masters in the art of land reclamation and irrigation. Caravans of llamas carried small skin sacks of earth to the steep valleys, where rich top soil was lacking. Here men built gigantic terraced walls faced with stone to hold the garden earth, which was spread on a specially prepared foundation.

Many slopes had more than fifty terraces, each ten feet high, rising like a huge staircase on either side of a valley. Such an engineering project was as sturdily built as the pyramids of Egypt. The Indians of today accept these terraces without realizing that they have not always been there, that laborers carrying loads of soil created every square foot of these "step" farms, which seem like the work of giants.

The Incas deliberately planned how to make each valley support a large population. Because of the high mountain ranges lying between the streams, each community had to be independent of the rest of the nation. It was also necessary to increase the growth of useful plants and destroy useless ones. Water had to be conserved, and river courses were straightened and held inside of walls. In many places, aqueducts led far back into the mountains, bringing moisture to valleys where streams were likely to be dry several years at a time. Inca engineers planned these carefully, providing manholes in the roof of the main conduit so that workmen could go down through the top and clear obstructions. Each family took its turn in receiving water for irrigation. Where the channels dropped down from one terrace to another, bathing places for field laborers were built into the rock walls.

The Incas understood agriculture better than most people. They fallowed the land and used fertilizer, most of which was obtained from guano found on the islands off the southern coast of the country.

Guano is bird manure, considered so valuable even at that early date that the Peruvians used it as an article of trade. The Inca government divided equally among certain villages the privilege of obtaining guano. It was unlawful for any man to land on the islands during the birds' breeding season or to kill them. Violators were punished by death.

Most of the guano was needed for the corn and pepper fields. Peppers and squashes generally were planted in the lower, moderately warm portions of the valleys, where the land sloped more gently.

More plants seem to have been domesticated in Peru than in any other place in the world. Quinoa, a grain somewhat like rice, was often grown with the corn. Four kinds of beans, including limas, were sown in the unirrigated areas, and potatoes, *ocas,* and *añus* in the poorer lands. The Andean Indians had a bewildering variety of edible roots, which, when seen in their markets, appear at first glance to be potatoes of many colors— yellow, red, brown, violet, and white. But among them are several kinds of sweet potatoes, the *oca* tubers, which are related to our sheep sorrel, and the *añu,* which belongs to the nasturtium family. There was another almond-flavored root called *ynchic,* so rarely found that poor people offered it as a gift to nobles. Another luxury was the *cuchuchu* rush. Then there were the various squashes, all sorts of tomatoes and peppers, and the lupin, which was used as a potherb.

Peruvian meals were mainly vegetarian. The common people had meat only when nobles presented it to them after hunting expeditions. The Indians dried these gifts of flesh to make "jerky" (known as *charqui*), so that it would keep for a long time. Their only other source of meat was guinea pigs, which they raised at home.

On a national feast day the Peruvians were usually sure of a special treat, for it was the custom to sacrifice llamas from the royal herds. After the religious ceremony the animals were taken to the public square and roasted. Then some of the meat was given to each family.

The rest of the year the farmers lived mainly on field herbs. Sometimes they went to stream beds and gathered cress or other plants and ate them fresh, as we eat radishes, or else boiled them. Some were dried in the sun to be kept for the season when there were no greens. Almost no salt was used in the food, but *oca* root, quinoa-leaf soups, and potato gruels were seasoned with capsicum peppers.

South and Central American farmers did such splendid work in domesticating peppers that plant breeders have not tried to improve upon them. The Indians at the time of Columbus already knew all of the kinds we use today and many more besides.

All the members of the *Capsicum* genus belong to the New World, and there were no names for them in the languages of Europe and Asia. When the discoverers came, they found the plants growing in the West Indies and Mexico. It is believed by some botanists that our favorite sweet variety was a native of Jamaica. When taken to Spain, it spread to other countries over the same road as that followed by corn. Spanish galleons also carried it to the Philippines and from there it went to

China. The Moors carried it to Africa and then east along the Mediterranean to Turkey, where it was immediately liked. For this reason capsicums were called Turkish peppers. As the Sultan Suleiman the Magnificent spread his conquests into Europe, his soldiers carried peppers to Hungary and there they are still the foundation of a national industry, the manufacture of paprika. The greatest pepper market in the world is in Szeged, Hungary.

Peppers were liked everywhere they were introduced, and in a short time their culture became almost worldwide. They were among the first products Columbus took home in 1493, and today there is no more typical sight in a Spanish market stall than gay strings of dried peppers.

27.

The Tomato

One of the first tastes today's American baby recognizes is that of tomato juice. A hundred years ago children were not so lucky; their mothers would just as soon have offered them poison. Tomatoes, they declared, were attractive to look at but were to be shunned as food. They were one of the last vegetables to be added to our diet.

Why was this red fruit so intensely disliked? And how did it prove itself?

The story goes back only a few hundred years. Tomatoes were never mentioned in books before 1554, when an Italian sketched and described the "golden apple" that had reached his country from Peru. The name by which we call it today was borrowed from the Mexican *xitomate*. Historians have little to say of it except to mention the red fruit eaten in South and Central America at the time of the Spanish conquest. It may first have been planted in Mexican maize fields, but because the tomato was perishable, it did not join the front rank of important foods. The Indians did not use it as a subject either for pictures or sculptures as they did corn, squashes, and beans. No traces of it were found in their tombs. However, offerings of tomatoes were made to the gods, and Inca and Maya women rubbed the red juice on their faces as a cosmetic.

When the plant was brought to Europe by Spanish explorers, its resemblance to the mandrake and the deadly nightshade was noticed. This was enough to brand it as poisonous. The potato and the eggplant fell under suspicion for the same reason.

Solanaceae, the nightshade family, to which all of these plants belong, is composed of both edible and inedible members. In ancient times it was associated with deadly potions, celebrated crimes, and witchcraft. A man or woman wishing to do away with an enemy went to a witch doctor or an alchemist and purchased a poison to put in the food or drink of his victim. Often henbane, nightshade, or mandrake, all of the Solanaceae family, were used for this purpose.

Scientists wonder if this family received its name from *sol,* meaning the "sun," or from *solor,* indicating narcotic effects. Not all species grow in the same manner, nor are the same portions of them edible. For instance, we eat the enlarged part of the underground stem of the potato plant and the fruit of the tomato and eggplant.

Botanists are unable to trace the tomato before it arrived in Europe on a Spanish ship. When the idea finally prevailed that tomatoes were not poisonous, a new belief was voiced—that they caused those who ate them to fall either in or out of love.

The Italians were the first to realize the food value of this "love apple" and they used it in cooking while English writers were still poking fun at the plant. One celebrated British gardener declared that if the red fruit possessed any nourishing qualities, they must certainly be very bad ones.

Nevertheless, by the seventeenth century Spaniards and Italians were boiling tomatoes with pepper, salt, and oil, or making sauces and soups from them. Elsewhere, tomatoes were considered merely ornamental plants.

Thomas Jefferson, probably influenced by his Italian friend, Dr. Philip Mazzei, grew some in Virginia in 1781, but Americans thought no more highly of them than most Europeans did. Within the next twenty years a Frenchman in Philadelphia and an Italian in Salem, Massachusetts, tried in vain to convince fellow townsmen that they could safely eat the fruit. But no, citizens of the United States would have nothing to do with it.

The French Revolution was the history-making event that changed people's opinions about the tomato. In 1783, when citizens of Paris were wearing red caps to show that they were republicans, a patriotic chef concluded that they ought to eat red food. He had heard that the Italians and Spanish found tomatoes palatable. He also knew that they were not an approved dish of the French aristocracy. Here was a chance to offer the revolutionists a symbolic food that had never been associated with the detested nobility. So he served stewed tomatoes to prove that he was an ardent republican.

Once the French gave it the stamp of approval, no praise of the new food was too lavish. People ate tomatoes cooked and in salads, in sauces and marmalades, and soon discovered medicinal virtues in the fruit. Doctors prescribed "extract of tomatoes" in the most extraordinary cases. They mashed the fruit in oil and placed it on bad bruises or rubbed it on the body to cure rheumatism. They used the juice for inflammation of the eyes and ears and in hot applications for various ills. In the French colonies in the West Indies, tomato pulp was placed on the eyelids to remedy ophthalmic troubles.

Early in the nineteenth century the reputation of the tomato underwent a quick change in the United States. One day in 1832 a boy who lived on an Ohio farm was walking along a country road when he saw a plant with bright-red fruit growing beside a rail fence. He gathered some and took it home.

"What are they, Mother?" he asked.

"You must not eat them, child," she answered. "They must be poison, for not even hogs will touch them."

"But what are they?" he persisted.

"Some call them Jerusalem apples," his mother explained. "Others say they are love apples. You may put them on the mantel to look at but you must not eat any."

The boy was curious about the fruits and kept his eyes open for more. He found some with purplish coloring and others of yellow tint, which he added to his collection. Naturally the tomatoes did not keep for many days, but long after they were gone the boy thought of the tempting-looking fruits. He must have ventured to taste them, for he remembered their tough pulp and how sour, hollow, and watery they were inside.

Ten years passed, and when the boy, grown to manhood, heard that other persons had become reckless and eaten the fruits with no ill effect, he vowed he would experiment with them. When he had a farm of his own, he planted a few tomatoes. They fascinated him, and he saved the seeds. He wanted to grow a smoother-skinned, better-tasting fruit. Each spring he sowed the seeds of his best plants separately, and after fifteen years of experimenting he was ready to give his own improved tomatoes to the seed trade. His name was A. W. Livingston, and he was among the growers who developed many of the best-liked varieties in use in the United States.

The earliest tomatoes must have been very small fruits, and a hundred years ago, tomato skins were still rough and uneven.

Not all tomatoes are red today; some are pink or yellow, and their shapes resemble cherries, pears, currants, or plums. They range in size from half an inch in diameter to six inches. In Ecuador one kind grows on woody bushes and others grow on a tangle of vines that cover the surrounding vegetation. In hothouses, ordinary tomato plants have been known to spread in much the same fashion, sometimes with runners forty feet long.

The plant is a perennial in the tropics, but in northern lands it must be forced under greenhouse glass if it is to bear the year around. It is both sensitive and hardy. Frost and flood will kill it, but it will stand much punishment of other kinds and will grow in poor soil if it has plenty of sunshine.

When the Spanish conquerors explored Mexico and Peru, they found native herb doctors curing diseases with tomatoes. They did not merely apply the fruit to the body, but, without knowing why, gave it as internal medicine. Four hundred years were to pass before scientists would discover the answer to this riddle. The patients for whom Indian wise men prescribed tomatoes were not getting enough vitamins in their food.

Vitamins are a mysterious "something" human bodies must have for healthy growth. Each of the several vitamins fills a need, and when we are deprived of these substances we become ill. Years ago doctors wondered why certain persons who had plenty of food lacked vigor and appeared to be in poor health. Then, about 1912, they realized that there is a hidden hunger, that men cannot be well unless their foods contain certain elements.

To learn which foods contained these elements, scientists experimented with animals. They fed cows nothing but oats; they gave starchy meals to rats. Then they added wheat and corn to the diet of the cows but omitted green alfalfa. They tried new combinations of fats, sugars, and salts on the rats. They stuffed all of the animals with food, but not one thrived.

Scientists knew that the answer to the puzzle must be near at hand. They were like treasure hunters piecing together clues on a map, but instead of digging in the earth they did their experimenting in laboratories with test tubes.

The discovery of vitamins was a gradual achievement, but the real breakthrough came in 1913, when Elmer V. McCollum and Marguerite Davis, at the University of Wisconsin, and Thomas B. Osborne and Lafayette B. Mendel, at Yale, working independently, discovered vitamin A, the anti-infection vitamin. The quest grew more exciting as scientific men the world over took up the search for more of these substances which are so vital to man's existence that they are named "vita," for life itself.

One after another, foods were discovered that contained these elements, and finally the laboratory experts were able to say, "We know what some of the vitamins are made of and can now give them to you in synthetic substances."

But no one who eats a well-balanced diet need buy his vitamins in a drugstore. Leafy greens, yellow and green vegetables, butter, milk, lean meat and fish, fruits, oatmeal porridge, and whole-wheat bread contain these precious materials.

Although vitamins are necessary to a person all his life, the greatest need is in his growing years. In order to become strong and healthy, children must have a balanced diet, with plenty of fresh vegetables and milk.

One of the greatest sources of vitamins is the tomato, which contains four of these magic substances. One is vitamin A, called the first line of defense against harmful bacteria. Another is vitamin B (thiamin), which aids digestion, gives one a good appetite, and keeps the nerves healthy. Lack of it caused the disease known as beriberi. Tomatoes also contain vitamin K, which helps coagulate the blood, and vitamin C (ascorbic acid), which cures scurvy, an illness common among people who do not get a sufficient amount of fresh foods. Although other vitamins may be stored for a time in one's body, fresh supplies of vitamin C are needed constantly. Explorers, seamen, and soldiers are the most frequent victims of scurvy. In time of war, or when families feel the pinch of poverty and eat mostly starchy foods, the disease creeps up on them almost unawares, loosening their teeth and making them listless. When a doctor examines a baby with bleeding gums and sore legs, he is apt ask why the child has had no tomato or orange juice to drink.

It was thought that God gave people scurvy in order to punish them for their sins, until medical men discovered that fresh foods would cure the disease. A British naval surgeon, James Lind, caused the adoption of lime juice as a scurvy preventive in the British navy in 1795. He insisted that the men drink lime juice as a preventive against this malady. When the Admiralty noted the good health of these sailors, it ordered that henceforth barrels of this health-giving juice be supplied to the navy. That is how English seamen got the nickname "limeys."

Even before this discovery, the importance of fresh foods had already been realized by Captain James Cook, who sailed in 1776 to explore the South Pacific, and completely upset existing ideas as to what mariners should eat. He had no wish to add to his own hardships or those of his men, and reasoned that if they were not healthy, he would never be able to accomplish the task before him. So from the beginning he determined to force them to eat what was good for them.

Since sailors in those days were used to living mostly on salt meat, oatmeal, rancid butter, beer, and rum when at sea, many of Cook's men rebelled at the notion of downing generous helpings of sauerkraut, fresh onions, and stewed grasses, which Cook ordered gathered and prepared for them when they stopped at islands. The crew first showed signs of rebellion at Madeira, and Cook saw that he must impress them with a necessity for obeying his peculiar diet rules. As an example he sentenced two men to twelve lashes apiece because they refused to eat fresh food. After that the others grumbled but ate what was put before them. Only after they had been at sea many months did they appreciate their captain's wisdom in insisting upon fresh vegetables.

Nowadays we have more pleasant methods of avoiding scurvy than by devouring raw onions. The explorer, the lone trader at an arctic outpost, the prospector far from stores, all can keep supplies of canned tomato juice and citrus juices to furnish the proper vitamins.

Cooking or canning does not destroy the magic power of the tomato. From the time that scientists determined the valuable qualities of the "love apple," it took an important place among foods, until now thousands of acres are devoted to its culture, and the huge industry that exists today produces catsup, sauces, and juice, and makes canned tomatoes available for home use in a variety of ways.

When you read of the elixers mixed by alchemists of old, you may be sure that none of them was endowed with such health-giving gifts, such delicate flavor, and such enticing color as the glass of refreshing tomato juice on your table.

28.

Carrots

Centuries ago doctors noticed that certain foods were good medicine. Even primitive man had a hunger for vitamins, which he did not understand but which somehow guided him in his choice of things to eat. The first mess of spring greens consumed after the long winter awakened his body and acted as a tonic. Being ignorant of science, he did not know that this was because his blood and tissues needed mineral salts and t digestion was stimulated by cellulose, or roughage

Medical men of the Greek, Roman, and Arab lizations had little knowledge of anatomy and r photographs to help them understand the hr functions. Often doctors did not know the v that were at hand, or how to utilize ther advantage. On the island of Crete, fc a lacy-leaved vegetable called *karot*

seeds and a yellow root. It had been brought from somewhere in western Asia. The islanders gathered the seeds
and used them for medicine (among other things, they
were considered a remedy for snakebite), but the people
paid no attention to the rest of the plant until some
hungry soul pulled one out of the ground, nibbled it, and
found that the herb had a crisp and tasty root.

Actually the plant had been used as food in India in
fairly early times, but it was so little known in the West
that its history is obscure. Wild carrots were native to
Europe but were regarded as a bad weed. There is a
dispute to this day as to whether the succulent-rooted
varieties were descended from the same stock as the wild
carrot.

Arabs in the tenth century were impressed with the
flavor of the domesticated carrot and carried improved
varieties to Persia, which may have been the very source
of their original wild stock, if indeed that stock was the
parent of our present delectable vegetable.

Carrots are related to parsley and about the only use
that had been found for the wild plant in the British
Isles up to the time of King James I, early in the seventeenth century, was as decoration for women's head-
dresses.

An agricultural writer from the Netherlands, visiting
Asia Minor and Syria in 1573, listed carrots among
kitchen-garden products he had seen.

Gradually after that, the domesticated carrot spread
to general use in European stew kettles. It was grown
and improved especially in France, Belgium, and Hol-

land and all the varieties we know today were developed there.

Although the carrot's first application was in medicine, its most important property was not known or appreciated. Only in this century (in 1919) have scientists discovered that it receives its bright tint from a yellow pigment called carotene. Wherever carotene occurs in foods one may be sure there is also vitamin A. Diseases caused by lack of this vitamin are rare in places where carrots, pumpkins, sweet potatoes, peas, cabbages, and other green and yellow foods abound. The vitamin A contained in animal products comes from the green and yellow substance of the plants that are fed the stock and poultry.

When people do not get enough of this vitamin, they suffer from night blindness. Medical men have known for many years that malnutrition affects the eyes. This fact was first brought to the attention of the world during David Livingstone's exploration of Africa. His men were suffering from night blindness caused by a lack of vitamin A, which preserves the pupil in the eye and enables people to see in dim light.

Today the poorest families need not lack this vitamin, for it is present in many inexpensive vegetables that are available throughout the year.

To get the most minerals and vitamins out of carrots, they should be eaten raw. Well-meaning cooks destroy the valuable properties of vegetables by cooking them too long or in too much water. Cooking causes vegetables to lose some of their calcium, iron, and phosphorus, which are drawn out in the juices. We need these minerals: phosphorus occurs in every living cell, calcium is the foundation of bones and teeth, and iron produces red blood cells.

The Chinese and Japanese are said to have excellent teeth because they eat so many green leaves. From the fact that the buffalo and other animals live solely on grass, we know that the leaves of certain plants constitute by themselves a complete food. Such a diet would not satisfy today's Americans or Europeans, but vegetables such as carrots are most beneficial when eaten raw, as in salads.

There was a day when salads were a luxury. In sixteenth-century England not even the queen could always have such a simple dish. Henry VIII's bride, Catherine of Aragon, once asked to have fresh salad with her dinner. But although the royal stewards searched far and wide, they could not find a single salad plant in season. Henry thereupon vowed this should never happen again.

It was in the end beneficial for England that Henry delighted in pampering his stomach. When he succeeded to the throne in 1509, the English people were living on a diet that was both dull and poor in vitamins, and they did not know how to grow anything better. The peasants of Merrie England were not as happy as storybooks would have us believe. They lived in drafty, unclean houses, and their food, besides being poor, was always scarce.

Eventually the farmers made a discovery. They had heard how much King Henry VIII enjoyed his food. Servants who had been at court gossiped about the monarch with the tiny mouth and the large stomach. They told how he sat at the dinner table stuffing meat into his mouth with a knife, how his eyes twinkled with happiness as he munched, how he grabbed restlessly from dish to dish. They described meats soaked in sauces of parsley, garlic, quince, pears, and wine, veal broiled with sage, and great pastries glittering with sugar or dotted with dates.

No vegetable or fruit was too rare for him to sample. If British farmers wished to grow choice dainties for the royal table, King Henry would reward them well. Had he not, since the incident of Queen Catherine's salad, sent to Flanders for plants and gardeners so that in the future it would not be necessary to bring vegetables from the Low Countries? He was determined to have the foods he liked.

Henry's vast appetite had wakened the farmers. From then on, the keeper of the royal purse had a busy time paying the market gardeners who tried to please the monarch by growing vegetables for his luxurious feasts. Here was a farmer with "archecokks" (artichokes), here was another with radishes, and here was a man with garden cress he had raised from seeds brought from Persia and Cyprus. And here was another with a basket of crisp Cos lettuce. And now humble husbandmen, who had thought themselves lucky with an acre of ground, a cow, and a few rows of cabbages, radishes, parsnips, marrows, and melons, wanted more land to plant with vegetables. Their wives worked in their kitchen gardens, cultivating herbs and roots for salad and sauce, to boil and butter, and "strewing herbs" to lay down in the house to keep it sweetly scented, for sanitary conditions were quite primitive in those days.

Many of the vegetables were entirely new or were known only to a few persons. In English botany books, 1548 stands out as the year midway in this transplanting period when many common vegetables were introduced in England. What the king had started, his daughter, Queen Elizabeth, carried on. She saw that agriculture was good for the country, and under her patronage gardening became fashionable among the rich and was highly respected as an occupation.

The first teachers of the English in the art of farming had been the Romans. Caesar's men found that the primitive Britons at the beginning of the Christian era took barely enough time off from fighting among themselves to cultivate their fields. They lived on milk, animal flesh, and a few roots. Barley was grown only for the rich and powerful leaders.

The invaders showed the Britons how to plant orchards and herb gardens, for which they furnished onion, turnip, parsley, and lettuce seeds. But Roman legionnaires were not the best instructors in agriculture. Their own mode of gardening was careless, and after a time, when their influence was no longer strongly felt, the Britons returned to their own crude ways. What gardens they had were neglected. But behind the high walls of monasteries, turnips, cabbage, and parsley still were cultivated for the abbot's table.

When the Norman kings crossed the English Channel and stormed Britain in 1066, they brought with them a few more green foods, such as fennel, broad beans, and field peas, but these additions to the meager supply of edible plants did not make vegetarians of the British. The whole nation was ill nourished, living mostly on salted foods, and the lack of fresh fruits and vegetables caused scurvy and skin diseases. Animals generally were killed in the autumn and the meat was salted down, for there was no fodder to keep them through the winter. A single scanty harvest brought hunger to whole districts. Entire families, driven from their homes by famine, drifted about, hoping to better their lot. Sometimes they starved to death on the road. Food was terribly scarce, and men ate wild herbs, the bark of trees, and uncooked meadow grass.

One day in the thirteenth century a noble lady, seeing peasants picking grass in the fields, asked her servant, "Can they not eat bacon and peas?"

"They could if they had it, my lady," he replied.

But conditions in the castles were not very much better than they were in the huts of the peasants. In the cold months the families of the nobles had hardly any vegetables, and the beautifully gowned ladies, shivering at their embroidery frames, would think of the pleasureable change of diet that spring would bring them. For all their formal manners and jeweled gowns of brocade, the nobles knew little of comfort. There was not so much as an easy chair to sit in or a fork to eat with.

As for food, the banquet table might have roast swans, peacocks, rabbits, venison and larks; it might have every kind of seafood from porpoise to minnow. It might have jellies and custards by the score, but seldom would it have fruits and vegetables.

The greatest baron in England breakfasted on bread, beer, wine, and either boiled beef or mutton. His ten-year-old son had half a loaf of bread, four pints of beer, and a chicken or boiled mutton bones. But this was no more extraordinary food for a growing child than the salt fish, sprats, and herring it was customary to feed to babies.

Although after the Norman Conquest rich nobles paid more attention to their kitchen gardens, the few cucumbers, garlic, leeks, and savory raised for the manor houses were unknown to the common people. Laborers lived on dark bread and salted herring, with maybe an egg as a special treat at harvest time. The men in the fields were thankful when they had cheese or strong onions with their dry loaf and jug of ale. Fresh milk meant a feast. Beans were often mixed with the barley or rye flour in the bread. Bacon and cabbage were signs of prosperity.

In the fourteenth century distress became so great that people willingly ate dog and horse meat when they could get it. Pets were not safe from thieves. A bishop was one day attacked by a famished crowd that demanded his fat horse and devoured it before his eyes.

The general health of the people would have been much better if they had eaten better foods. Medicine went backward in the Middle Ages. Some doctors believed that the stomach was a caldron in which food was cooked by the heat of the liver.

The fifteenth century brought more beans and bacon for the poor British farmer, a few onions and leeks, and, in the lean months, pea or bean loaves or pottage. Then there came a sad time when famine and plagues and the Wars of the Roses swept through the country. Things changed after that. Workmen demanded more to eat because these disasters, which had depopulated the nation, taught them that something must be done to better their manner of living.

Events that the common people did not comprehend were taking place elsewhere in the world. America was discovered and the precious metals of Peru and Mexico were pouring into Europe. This changed the value of money; it would not buy as much as formerly. People could not understand why food prices were high. They said it must be because the king had turned the monks out of the monasteries, where the poor had always gone in times of need to ask for food.

In their own villages life was also different. During the wars many spendthrift nobles had been forced to sell their estates, and these had been broken up into small farms. Where peasants had been content to grow only what was needed to feed their village or manor house, they now wanted to raise extra crops for market. They showed a stronger instinct for trading and spoke of the nest egg they wanted to put aside. They no longer left their fields wide open but fenced them with hedges to keep out weeds and stray cattle and to prevent confusion of ownership during the harvest.

With the monasteries ceasing to be like great feudal walled fortresses, some of the garden lore that had been guarded within them leaked out. If monks could grow vegetables, why not the farmers? And if they could do this, so the farmers told themselves, their earnings should go into their own pockets, not into the treasury of an overlord.

The art of vegetable gardening came to England from Holland. Gardeners from Flanders and the Netherlands went to the county of Surrey to sow cabbage, cauliflower, turnips, parsnips, French peas, horseradish, and lettuce. One of the plants the Flemings brought was the carrot, which thrived especially well in the soil of Kent. Beets, potatoes, and scallions were added to the list, and even the turnip seemed like a new food because it had been completely forgotten during the Dark Ages.

Salted-meat and bean-bread dinners were a thing of the past. Vegetables that formerly had been luxuries of the rich gradually became common dishes among the poor.

Although the British population has greatly increased since King Henry's reign, today the poorest peasants are better nourished than were the great overlords of his day. The English countryside is a land of kitchen gardens, and in some of them may be found growing a species of spinach named Good-King-Henry in honor of the merry monarch whose love of food changed the diet of a nation.

29.

Parsley and Celery

The heroes of whom the poet Homer sang were simple eaters. After performing deeds of valor, they sat down to dinners of bean or pea pottage, salads, cheese, figs, and bread dipped in wine and water. One hero washed cabbages and ate them fresh. Another peeled onions at the table. But no one classed parsley with these foods, for its delicate foliage was an emblem both or bravery and of death. It was fed to the chargers of great warriors before they went into battle, and it was made into crowns for the victorious champions when they returned.

Hercules, the Greek hero known for his great strength and courage, is often pictured as wearing a parsley wreath, a tribute to his valor. Youths who won marathon races or threw the discus or javelin farthest were decorated with parsley crowns. These were also given honored guests, for the odor of the herb was said to excite the imagination.

The Greeks had another and less cheerful use for parsley and celery; both of them were associated with funerals. Tombs were decorated with the wreaths, and when a person was dangerously ill his friends sometimes murmured, "Alas, he now has need of nothing but parsley," meaning he will soon be dead.

Today, parsley is used as seasoning in many culinary dishes, and cooks often do not realize that it has great nutritive value. It is one of the three foods richest in vitamins A and C and an excellent source of vitamin B_2 (riboflavin), as well as being tremendously high in iron. Chefs who put sprigs of parsley on plates of food do not do this for decorative purposes alone, but also for good nutrition. That spray of green should not be pushed aside, as is done by so many diners, but should be eaten.

One interesting story of the vegetable kingdom concerns the way in which man has changed the plants he eats. Nothing illustrates this more clearly than celery. The crisp white stalks of today did not exist a few hundred years ago.

Although celery has been known for three thousand years in Europe, eastern Asia, and North Africa, it was in ancient times a kind of wild parsley used in wreath-making. Its leaves were much as they are today, but the stalk was short and unattractive, like that of the turnip-rooted celery, or celeriac. The Greeks called the plant *selinon*, and named a city for it, Selinon. They used a picture of the leaf on coins.

The only other use the Greeks had found for celery

was for medicine. In faraway China, however, people ate the short, stubby stalks, and Arab merchants going back and forth may have conveyed to their countrymen that this plant was worth growing as food. It had reached the gardens of Syria when the lowland farmers of Italy set out the first European plantings in the sixteenth century. Gradually the wild plant, which had been thought to have poisonous properties, was cultivated for the table. The stalks, known throughout the Middle Ages as *selleri,* had not yet been endowed with the sweetness, crispness, and appetizing qualities now associated with them.

There are two kinds of celery in our markets, one desired for its fleshy root, the other for its leaf stem. *Celeriac,* grown for its root alone, shows another change man has made in the plant under cultivation, for it is very unlike the original wild celery.

Crisp white celery stalks were given to us by modern plant breeders. Until recent times good celery was a luxury. People grew it, but the stems were pithy, slightly bitter, and dingy-looking. English and American farmers discovered that if they shielded from light the lower portions of the plants by covering them, the stalks would be whitish and much better in flavor. For that reason it is now customary to bank the plants with soil as they stand in rows, or place planks against them so that only the tips of the leaves are exposed to the light. This procedure is called blanching.

Pithiness in celery, breeders learned, was caused by collapse of the cells, an inherited trait of the plant. Therefore, care was taken not to mix good seed stock with that from pithy ancestors. Another improvement made the plant resistant to a disease called "yellows," which spoils the color and stunts growth.

The discovery that both celery and lettuce are foods loaded with vitamins made these vegetables so popular that today they are the largest salad crops in the United States. Many millions of dollars' worth of celery is raised in the United States every year, most of it in California, Oregon, New Jersey, Florida, Colorado, Michigan, and New York, where the proper dampness is present in the soil.

Americans wanted to eat the crisp stalks in every season, so breeders developed some types that mature earlier than others. Chinese gardeners living around Great Salt Lake in Utah are credited with giving the world a new winter variety whose stems have no strings.

30.

More and Better Vegetables

How was man able to bring about changes in food plants? Where did we get crisp heading lettuce and coreless carrots if Nature did not give them to us?

Many varieties of vegetables in seed catalogs have become available only in recent years. They are the result of the work of plant breeders both in this country and abroad. Plant breeders greatly depend on foreign seeds and cuttings to further their work.

The introduction of foreign plants in America began the day Columbus instructed his men to sow European garden seeds at Isabela, Santo Domingo, the first Spanish settlement in the New World. New plantings were carried on intensively in the colonial era, but at that time the purpose was mainly to grow food for the hardy adventurers who pioneered the way.

The conquest of the New World was to a large extent

prompted by a desire for improvements in agriculture. Only a few years before Columbus made his first voyage, Spain had still been occupied by the Moors, then the world's greatest farmers. But the lessons they had taught bore fruit long after they were driven back to Africa. The Spanish colonies did not suffer any such period of hunger as was endured by the colonies of Great Britain. Wheat, sugarcane, seeds, plants, farm laborers, gardeners, tools, and irrigation were almost as important to the Spanish conquest as armaments and soldiers.

Hernando Cortez, who penetrated Mexico in the early sixteenth century, wrote a letter to the king of Spain begging him to permit no ship to sail for the West Indies without bringing plants and seeds in its cargo. This request soon became a law. Cortez saw that the Mexicans were skilled cultivators, and he promised the king that New Spain soon would produce a great abundance, so that some of the yield could be sent abroad.

Before all of the thirteen American colonies had been founded, the Latin American countries already were exporting agricultural products that had not been known in the Western hemisphere until the conquest. The introduction of European vegetables into Peru was rapid, partly because the Spaniards enjoyed being surrounded by familiar plants. No intelligent colonist made a trip to his homeland without bringing back some plants from the old country.

The impression that the new foods made on the Peruvians was remarkable. It was a great occasion in an Inca nobleman's house when the Spanish treasurer sent three stalks of asparagus or a tiny bunch of carrots as a gift. The Inca invited friends to witness the cooking of these new vegetables over the coals in a brazier placed in the main hall of his home. With his own hands the host cut the carrots in small pieces and offered to each guest a morsel of the curious food of the Spaniards.

Within twenty years this same nobleman's children were eating the new vegetables as though they had always grown on the Andean valley terraces. The Peruvians soon were enjoying more kinds of vegetables than any other people, for in addition to those native to their country, they had the common ones that the Moors had given to Spain.

Root crops were not as well received in Peru as the more unusual vegetables, such as globe artichokes and asparagus. The Indians scorned turnips because there were so many other roots they already knew, but this vegetable filled an odd need in the province of Quito. It was grown for turnip-seed oil, used in preparing wool for certain types of cloth. In their native land the Spaniards would have used olive oil for this purpose.

Cod fishermen and sailors in the North Atlantic introduced the first new plants on North American shores. Out of curiosity they sowed little patches of seeds to learn whether they could be made to grow on the new continent. Indians, finding these plantings, made good use of them.

Sir Walter Raleigh's settlers in the spring of 1586 planted the first real crops sown in the United States, but these colonists did not stay to reap a harvest. Two weeks before the barley, peas, and oats were ripe they returned to England with Drake. Another expedition sowed garden vegetables in 1603, and ten years later more than a dozen European food plants were growing in Virginia. Naturally each group of colonists brought more vegetables. Even before the Revolution, American patriots believed they should try to grow a variety of useful plants so as not to have to send abroad for products that could be raised successfully on their own soil.

Benjamin Franklin, Thomas Jefferson, and many another early statesman helped introduce foreign plants into the country. George Washington, himself a farmer, was the first to suggest that a branch of the government should be devoted to agriculture. Partly because voluntary gifts of foreign seeds and plants had been sent or brought home by Americans who made trips abroad, Congress in 1839 set aside a small sum for their care and distribution. This was the beginning of the Department of Agriculture, but until 1889 it was not considered sufficiently important to have a seat in the President's cabinet.

Toward the end of the nineteenth century the department expanded and was divided into many sections, including one on plant pathology, which engaged in the study of plant diseases. This work is now done in branches of the Agricultural Research Service.

Illness and enemies come to plants as they do to human beings. But the diseases of humans can be treated in hospitals and their enemies can be sent to prison. The vegetable gardener has no such recourse. His seeds are sprouting beautifully in neat rows. Then one morning he finds the young plants blackened by an early frost, chewed by hungry slugs and worms, or covered with a disfiguring scale. It is the duty of plant researchers and plant breeders to help the farmer overcome such misfortunes. The hardy vegetables we have today did not just happen. They are the result of combining the enduring qualities of one variety with the tasty, useful qualities of another.

When some new pest comes to the attention of the Department of Agriculture, the word goes out to its men in the field to look for a tomato that is resistant to wilt, or a potato that will stand up under frost. Hundreds of packets of seeds begin to come in, not from plant explorers alone, but from agencies and individuals abroad who correspond with the Crops Research Division, which through its New Crops Research Branch has constant sources of aid in other countries.

As each parcel of seeds or cuttings arrives, it must be disinfected or fumigated at the inspection house in Washington, to make sure that no undesirable plant diseases or insects are brought into the United States. After this the parcel often goes to Glenn Dale, Maryland, for quarantine detention. Cuttings must be grown at the quarantine station for a full year, but most seeds are distributed after fumigation.

As soon as the packets arrive from quarantine, the New Crops Research Branch gives each a permanent number and divides the seeds into lots. Then they are sent to different state or government plant experiment stations to be tested for various kinds of tolerance and for resistance to such things as poor soil, cold, insect pests, fungi, and bacteria. When a certain lot is found to have desirable qualities, it is sent back to Washington. There it undergoes the second stage of development, the breeding of hybrid or new varieties by combining the imported plant with the best standard garden types.

Vegetable seeds to be thus used are sent to Beltsville, Maryland, where in 1932 the United States government established the world's largest horticultural experiment laboratory. It has eight large greenhouses devoted to vegetables alone, and many times that number for other economic plants.

This vast laboratory does not emphasize the use of chemicals for the control of pests, but concentrates on the study of plant traits inherited through minute living elements called genes. By borrowing a color gene from one individual, a flavor gene from another, a health gene from a third, scientists can form a new pattern for a modernized vegetable combining the virtues of several gathered in widely separated parts of the world. The men who carry on this work are more than mere gardeners; they are highly trained technical authorities with doctoral degrees from universities. Their tools are camel's hair brushes, tweezers, paper bags, and cloth cages, and their plantings, dotted with odd, tightly tied hoods placed over blossoms, are curious to look upon.

Because some vegetables, such as turnips, rutabagas, and radishes, have nectar-bearing flowers which insects usually pollinate, all crawling and flying creatures must be kept away from the blooms the breeder intends to use for his work. Instead of permitting pollen to come from any nearby plant, he selects his sources and applies the yellow dust with a fine brush or shakes the male flower until it sheds pollen over the stigma of the female blossom. He then encloses the female flower in a transparent bag or a cloth cage, tied tightly at the base so that no insect can enter. It is here that the improved seed of the hybrid is born.

No two combinations produce the same hybrid, but the identity of the parents can be traced through their numbers, so that the laboratory always is informed which matings of plants have produced the desired results.

To understand the process better let us follow the travels of a tiny, hairy tomato sent from Peru. The explorer who found it near Lima thought that although it had no food value, it seemed a particularly sturdy specimen. After being tried out in various parts of the United States, it was crossed with one of the best American tomatoes. Pollen removed from field-grown plants of this useless-looking, catnip-smelling tomato was delicately brushed on blossoms of tried and trusted plants growing in a screened greenhouse. Immediately afterward, transparent bags were put over the plants and remained there until the fruits were partially developed.

The seeds of these fruits were collected and plants grown from them. This time they were planted adjacent to rows of wild and cultivated tomatoes having certain diseases. Some were planted in earth infected with another disease. At the end of the growing season, the hybrid plants were still healthy. The hairy tomato, almost useless for food, had given wonderful disease resistance to a delicious American table variety. In the past, the discovery of a wild disease-resistant plant was worth so much to farmers that if the plant explorer who found it had brought back no other specimen, it alone would have paid many times the cost of his long journey. It helped accomplish the aim of modern vegetable breeding, which is to produce hybrids with better constitutions than those of the food plants we already have.

Cross-pollination is not always the simple matter of transferring the yellow dust on a finely pointed brush. Some vegetables offer difficult problems. The pollen of beets is so fine that it is almost always present in air currents over the garden, so the female flower must be removed to the greenhouse or some place where there is not the slightest breeze when pollination is done. Carrot flowers are so small and difficult to handle that experts developed a special technique for pollinating them. The same method is also successful with onions. Its most important equipment is a quantity of baby houseflies, so young that they have not yet touched another object.

In order to have plenty of these freshly hatched pollen carriers, scientists must raise them in special cages. Eggs of the flies are collected and the larvae are reared on a mixture of wheat bran, molasses, and yeast. As soon as the first pupae start to emerge from their cases, the entire lot is placed in cold storage, from which they can be removed as needed. They will continue to develop there but in the cold atmosphere will not leave their cases.

A few minutes in the warm air causes the pupae to show signs of life, and they are immediately ready for use when the carrot to be pollinated is in the right stage. The blossom of the pollen-parent is cut and

placed in a jar of water set so close to the growing female bloom that both can be covered with the same
bag. The pupae are placed inside the transparent sack,
which is securely fastened at the base. As soon as the young
flies emerge from the pupa cases, they go to work,
feeding on the nectar of the blossoms and meanwhile
carrying the pollen back and forth on the hairy coating
of their legs and bodies. They pollinate the flower a
hundred times a day; man, with his clumsier artificial
methods, might do so only once. The flies live three or
four days, then are replaced with others.

By such painstaking methods man is developing vegetables to suit his needs, altering not only their form
and physical stamina but giving them additional nutritive content. An example of the manner in which man
has impressed his own needs upon plants is the sugar
beet, to which has been given a greatly increased ability
to produce sucrose.

In the course of searching for cold-resistant fruits and
cereals suitable for planting on the Great Plains at the
end of the last century, the Bureau of Plant Industry
sent several explorers abroad. These men collected many
useful specimens in addition to those they were instructed
to seek. David Fairchild, one of the pathologists, con-
ceived the idea of a new office to take care of this widespread work, and in 1898 the Section of Foreign Seeds
and Plant Introduction came into existence. Today the
work is an important part of the vast machinery of the
Crops Research Division. In a little over forty years
the department has introduced more than 200,000 kinds
of seeds and plants into the United States.

David Fairchild was one of the men in the department who chose to follow the colorful career of plant explorer —a career less glamorous today than it was in his time. In an average year usually three such experts are working in various parts of the world. Although they seek mainly economic plants, they are always on the lookout for extremely hardy varieties of vegetables. Turkey and India are two of the main fields. Explorers also search South America for potatoes and tomatoes. The investigators often are able to find familiar old vegetables in their wild state, or the seeds of a certain kind that has thrived in one locality for hundreds of years under adverse conditions, which often bring about important changes in plants. Such specimens may represent the survival of a species and therefore have special qualities of resistance to attacks of disease and insects.

Early plant explorers had to visualize in which ways the varieties they sent home might contribute toward production of strains superior to those already grown. They also had to picture in their minds the regions in the United States to which these varieties might be best suited. Nowadays this is all planned before the explorers leave. They travel on a schedule, seeking definite items. The explorer's first concern is that his peculiar harvest of seeds, fruits, and living cuttings reaches the United States in satisfactory condition; that is, with the germ of life still in them. Sometimes in the past a small caravan was required to carry bales of moss and sacks for packing specimens brought out of Japan, Nepal, Bhutan, Borneo, Taiwan, Turkestan, or the Caucasus.

Although there is less glamour in the work today than

in the days when David Fairchild went abroad at the beginning of this century, unexpected triumphs still fall to the lot of the explorers. One of them was permitted to enter a region in Iran that had been closed to foreigners because of a native uprising. The explorer's quest was directed toward the study of pistachios, but seeing in a market in Kashan a small, tight-necked onion, he made a trip into the valley where it grew to obtain bulbs. He knew that the bureau was on the lookout for an onion to help California growers fight onion thrip. The tight collar and glossy leaf on the Kashan specimen were exactly what they needed to defeat the insect, and several years later this onion proved to be an important find.

The plant explorer, arriving in the district where he is to work, generally first pays a visit to a village market, where very often among the hodgepodge of articles offered for sale he gets the clue to something he is seeking. Very often his hunch comes from words casually dropped in conversation or from the sight of some unusual wild plant a peasant has found and ventured to bring to market with his produce.

Perhaps the explorer has arrived by automobile over a trail intended for donkey and camel caravans. He pulls into a village huddling in a valley and strolls with his interpreter toward the bazaar.

"Who is the most important dealer in seeds?" he asks.

Someone points the way to a dark shop, crowded with dusty packets, sacks, and drawers. The explorer does not plunge directly into his business. He is a foreigner and the dealer must first be assured that his customer's purpose is actually to buy seeds. They sip glasses of tea or small cups of sweet coffee. They never talk about prices.

The explorer asks, "Have you any eggplant?"

"How many do you wish?"

The explorer does not mention ounces or pounds. He cups his hands to indicate as many seeds as will fill his palm.

"How many kinds have you?" he asks the merchant.

"Ten," replies the bearded dealer, but when he has brought out his parcels, the number is close to forty. He speaks of their uses; this one is large and coarse; that small one is best stuffed with ground meat; here is a long, curving, white kind, and here is a greenish-white one.

The explorer makes his selections and the merchant pours a small mound of seeds on a square of old newspaper and deftly twists it into a cornucopia closed tightly at the top. The business with the merchant is completed, but the explorer lingers in the vegetable market. He pauses before a basket of eggplant and admires them.

"They came from my grandmother's farm," the owner of the basket explains.

"We should like to see the beautiful garden which produced them," the explorer's interpreter flatteringly suggests.

"It is ten miles away in the next valley," is the reply.

"We do not mind the distance."

The interpreter asks directions. The explorer is following one of his hunches. The eggplant is not really what has interested him, but in the basket he has seen a splendid tomato. He will drive over and inspect the old woman's garden, and maybe she will tell him where she obtained the seeds that produced the tomato. So the explorer is off on a fresh trail he might never have found had it not been for the peddler's basket.

New techniques of production, handling, and processing create constant need for improving vegetables. Canners some years ago asked for an ear of corn so slender that several would slip into a can. Seedsmen immediately went to work to supply this, so that "corn on the cob" could be eaten the year around.

Freezing has solved such needs and presented other economic puzzles. New vegetable products are constantly being developed, and the processor must have peas, carrots, or broccoli that will keep their color and good taste when frozen.

Scarcity of hand labor is also a reason for changes in vegetables. Take the tomatoes that are bred for machine harvesting. California's tomatoes used to be picked by migrant Mexicans, but the government no longer permits them to cross the border for this purpose. It looked as though some of the canners might have to move their factories to Mexico, but just about then a mechanical picker was perfected that cuts the plants underground, pulls them up with metal fingers, and gently shakes off the fruit for a crew of packers to sort and care for. This called for the breeding of a new tomato plant with fruit that all ripened at the same time and had easy-snap stems for bruiseless picking. Skins and interiors had to be sturdier than other kinds to withstand machine handling. Even the planting of these miracle tomatoes is different. The seeds are in plastic tapes spaced at intervals. They are attached to a planting machine and threaded down through a digging tube, so that one man can plant as much as thirty acres a day.

Now there are special machines for harvesting the leading vegetable crops. A man can ride in an air-conditioned cab and harvest a crop of corn that formerly required a crew of eighty. A species of factory-on-wheels moves down the rows on celery farms, cutting, trimming, washing, and crating at the same time, accomplishing what used to take forty men to do. "Old McDonald's farm" is a weird new world of robots, and all the desired changes haven't yet been brought about; farmers are asking for machines that will select as well as pick a product.

Man's cleverness has given us a supply of plant materials from the entire world and adapted them to the needs of modern life. This privilege is being extended to nations farther north, for breeders have developed strains of some vegetables suited to the short season in the Arctic. Weather and poor soil no longer limit farmers in their activities, for potatoes, tomatoes, and other vegetables can be grown successfully without earth, in shallow tanks and glass jars containing chemical compounds in water.

What men can accomplish was demonstrated when a Pacific clipper plane base was established at barren Wake Island, where the soil was so filled with calcium that plants ordinarily could not grow in it. By means of water-culture tanks the station was stocked with fresh supplies of radishes, lettuce, beans, carrots, squash, corn, cucumbers—altogether more than thirty varieties of vegetables where absolutely none had existed.

The average gardener is reluctant to grow unfamiliar plants. Many practical and delicious vegetables have been introduced into this country, only to be forgotten because they were not foods about which we read in our daily newspapers or hear mentioned on television programs. Like the broccoli that thrived as a weed on a Long Island refuse dump, unknown delicacies have their ups and downs. We hear of tasty chayotes, pear-shaped squashes that grow on arbors in the West Indies and are brought to New York from Puerto Rico to whet winter appetites of gourmets. Yet chayotes have been grown and nearly forgotten in Georgia because no one attempted to make them popular. Children of neighboring Mexico spend their copper coins for delicious slices of jicama, the turnip-shaped root of the yam bean vine, which is totally unknown in American markets. Chinatown shops of San Francisco are filled with crisp, sweet tubers of the water chestnut and other aquatic plants, but we have not yet attempted to sow vegetable beds in shallow ponds.

If some of us desire curious or exotic foods, for the time being we shall probably continue to depend mainly on airplanes to bring them to our tables from the tropics.

A few years ago many of the fresh foods we eat throughout the winter months could be had only during their short seasons. Lettuce was almost unknown in the western cattle country, and many families depended largely on canned vegetables. Refrigeration, frozen-food processing, and rapid transportation have changed all this. Household refrigerators have helped to increase the demand for many fresh and frozen foods.

Just as the tin can changed life a few generations ago by making it possible to preserve most of the harvest, so refrigeration again revolutionized this nation's eating habits. We no longer wait for spring to bring the first shoots of green vegetables. Every day of the year we expect food stores to offer us the choicest products of the nation's gardens. We are more demanding in our requirements—the lettuce must be in tight heads, the broccoli neatly bunched, the tomato evenly formed and colored. A thousand or even a hundred years ago the housewife would have wondered that we risked our lives by eating such plants. Today we dare not omit them from our diet.

Vegetables have come a long way!

Index

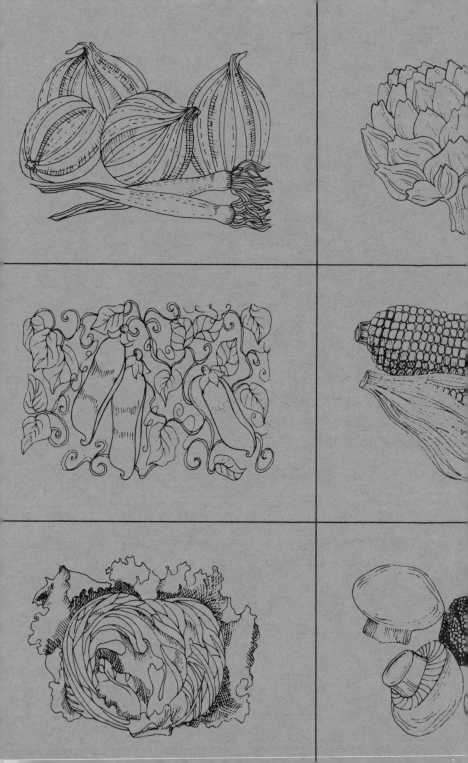